MW01040194

COVER UP

COVER UP:

What you are not supposed to know about NUCLEAR POWER

by
karl grossman

THE PERMANENT PRESS
Sagaponack, New York 11962

International Standard Book Number: 0-932966-10-1
Library of Congress Catalog Card Number: 80-81394
The Permanent Press, Sagaponack, N.Y. 11962
Printed in the United States of America

For Paul Jacobs

"I have set before you life and death, blessing and curse. Therefore choose life, that you and your descendants may live."

Deuteronomy

TABLE OF CONTENTS

INTRODUCTION

You have not been informed about nuclear power. You have not been told. And that has been done on purpose. Keeping the public in the dark was deemed necessary by the promoters of nuclear power if it was to succeed. Those in government, science and private industry who have been pushing nuclear power realized that if people were given the facts, if they knew the consequences of nuclear power, they would not stand for it. If people knew that the kind of accidents that happened at Three Mile Island, at the Fermi Reactor, at Browns Ferry, at Windscale, at "SL-1," among others—the sort of huge catastrophes which have been only barely avoided—are to be expected, they'd be damned upset and would insist a stop be put to nuclear power.

So an army of public relations practitioners has been working for decades to, in the jargon of the trade, make the people think of "Citizen Atom" as a friend, before the truth became manifest.

The "nuclear runaways" and "meltdowns," the "China syndrome" would come, it was figured. So would the "routine" releases of radiation from nuclear plants—and their results: cancer, leukemia and genetic injury. Also to be expected were illness, injury and death in connection with mining, milling, fuel fabrication, transportation, reactor operation and storing waste—the basic steps in the entire nuclear "cycle."

The U.S. government, when it considers whether nuclear plants should be built, is not ignorant of the costs. It submits them to mathematical analyses, postulates "deaths per gigawatt," even puts a dollar figure on your getting cancer or dying as a result of nuclear power, in order to establish a "cost-benefit" ratio. The government categorizes a series of accident possibilities as ranging from "Class 1" to "Class 9" catastrophic events. Class 9 is seen as potentially killing tens of thousands, causing cancer and genetic damage in many others, and costing billions of dollars in property damage. Nor are survivors assured of compensation. Because of a federal law designed to promote nuclear power, the Price-Anderson Act, Americans are not covered by their insurance for nuclear accidents.

Here is the "nuclear clause" of two typical insurance policies.

ALLSTATE INSURANCE COMPANY

9. Nuclear Clause: This policy does not cover loss or damage caused by nuclear reaction or nuclear radiation or radioactive contamination, all whether directly or indirectly resulting from an insured peril under this policy.

U9007 PRINTED IN U.S.A.

:ORMS AND ENDORSEMENTS, IF ANY, ISSUED
RED STANDARD HOMEOWNER'S POLICY.

INSURANCE FROM

CNA CNA Plaza
Chicago, Illinois 60685

INSURANCE IS PROVIDED BY THE COMPANY DESIGNATED BELOW
(A stock insurance company, herein called the company)

- [ ] **Continental Casualty Company**
- [ ] **National Fire Insurance Company of Hartford**
- [ ] **American Casualty Company of Reading, Pa.**
- [ ] **Transportation Insurance Company**
- [ ] **Transcontinental Insurance Company**
- [x] **Valley Forge Insurance Company**

6. Nuclear Clause: This policy does not cover loss or damage caused by nuclear reaction or nuclear radiation or radioactive contamination, all whether directly or indirectly resulting from an insured peril under this policy.

They are the same. They will match your insurance policy. Check it. And ask yourself: if nuclear power is safe, why won't insurance companies insure Americans against its hazards?

When all the deaths and cancer and mutations began happening, the nuclear promoters strategized, it would be too late. By then there'd be a dependence on nuclear power. And the fall-back PR line would be used: well, 50,000 people get killed on U.S. roads in auto accidents each year—are we to outlaw cars?

There's no comparison. We have a clear choice now–and not for much longer–to avoid the admittedly lethal nuclear energy highway, and, instead, follow the path of safe, renewable energy sources. We do not need nuclear power. We do not need to take a colossal risk that threatens our very survival, and the survival of those who are to come after us.

For nothing less than survival is at stake.

As three top nuclear engineering supervisors who resigned from the General Electric Corporation as a matter of conscience told the U.S. Congress:

> We did so because we could no longer justify devoting our life energies to the continued development and expansion of nuclear fission power--a system we believe to be so dangerous that it now threatens the very existence of life on this planet.
>
> We could no longer rationalize away the fact that our daily labor would result in a radioactive legacy for our children and grandchildren for hundreds of thousands of years. We could no longer resolve our continued participation in an industry which will depend upon the production of vast amounts of plutonium, a material known to cause cancer and produce genetic effects, and which facilitates the continued proliferation of atomic weapons throughout the world.*

Why choose catastrophe when there is an abundance of practical, cheaper, more job-producing energy alternatives: solar power, wind power, geothermal energy, power from solid waste, co-generation, energy efficiency, biomass power including agriculturally-grown alcohol (plant power) to propel vehicles and power from the tides and the waves? The list is endless. Nor is technology to be discarded. Technology can make photovoltaic cells which directly convert the energy of the sun into electricity, and can harvest wind power, water power and other natural bounties.

However, the oil monopolies–which now call themselves energy

*From testimony of Dale G. Bridenbaugh, Richard B. Hubbard and Gregory C. Minor before the Joint Committee on Atomic Energy.

companies—and the electric companies don't want any of this. It would mean that people would be able to be free of their domination. The sun and wind send no bills. So they, and the government they so easily bend, push us onward, forever manipulating.

A few weeks after Three Mile Island, the group which has become a kind of Board of Directors over governments in America, Europe and Japan—the Rockefeller family-led Trilateral Commission—held a conference. Having "our leaders" now "shift the debate from the safety issue to the energy supply issue" was stressed according to a commission bulletin. Lo and behold, a gasoline shortage and gas lines followed and America is told by Jimmy Carter, whose political base comes from The Trilateral Commission far more than it does Georgia, and who claims to be a nuclear engineer: "Nuclear power must play an important role."

In this crisis atmosphere, we are being told to swallow the nuclear pill. And you, on an issue on which your very survival hinges, have not been given the facts. For years the dangers and consequences of nuclear power have been obfuscated and suppressed.

Those in the media and in science making inquiry have had systematic attempts made to silence them. The political and communications processes have been perverted, neutralized in this great effort to modify society.

Have you ever seen this report? Ever read about it? Heard of it?

Federal Response Plan for Peacetime Nuclear Emergencies

(Interim Guidance)

April 1977

General Services Administration

Federal Preparedness Agency

4. Category III incidents.

a. Description.

These are situations in which, despite all preventive, protective and response efforts, an actual nuclear detonation or widespread radioactive contamination, shall have occurred within the United States.

The nuclear detonation could range from a very low to a very high explosive yield. Widespread contamination could come from such nuclear detonations, or it could conceivably come from other sources such as the sabotage of nuclear power plants or other fixed nuclear facilities, a serious accident involving the transportation of nuclear materials or nuclear power plants or other fixed nuclear facilities, or the explosion of a crude nuclear device resulting in the dissemination of radioactive material.

---

5. Category IV conditions.

a. Description.

A nuclear detonation or widespread dispersal of radioactive material can be expected to create, in addition to the need for immediate lifesaving actions and other related operations as described in Category III above, the need for long-range recovery and rehabilitation measures directed toward the permanent rebuilding and reconstitution of the socioeconomic structure, the physical facilities and institutions of the affected area(s) and the long-term reduction or elimination of radioactive contamination. These measures would involve such things as housing, utilities, hospitals, schools, business and financial enterprises, governmental structures, and organizations. These measures can be expected to continue for months or years after the immediate lifesaving operations have been completed, and should be administered by an organizational mechanism responsive to these long-range needs. Although it is difficult to provide any specific planning guidance for the long-range needs of an area affected by a peacetime nuclear emergency, it is essential that some forethought and consideration be given to such things as the reconstitution of local government operations, the rebuilding of the social and economic structure of the affected area(s), and the allocation of critical resources which may be in short supply following a nuclear detonation or dispersion of radioactive material.

DEPARTMENT OF HEALTH, EDUCATION, AND WELFARE is responsible for:

-- Assistance to State governments in the development of plans for the prevention of adverse effects from exposure to radiation, including the use of prophylactic drugs to reduce radiation dose to specific organs, and health and medical care responses to radiological incidents;

-- Issuance of guidance on appropriate planning actions necessary for evaluating and preventing radioactive contamination of foods and animal feeds, and the control and use of such products should they become contaminated;

-- Issuance of guidance on emergency radiation doses related to the health and safety of ambulance services, hospital, and other health care personnel, in cooperation with EPA;

-- Establishment and issuance of guidelines for radiation detection and measurement systems for use by ambulance services and hospital emergency departments, in cooperation with NRC; and

-- Provision of advice, guidance, technical expertise and materials, and financial assistance, if authorized, to affected State and local governments. This assistance is used to provide emergency medical services, public health measures, and rehabilitation services.

In addition to supporting State and local government activities, DHEW provides the following assistance directly from its headquarters or regional offices or through the detail of personnel to other Federal, State and local government agencies:

-- Evaluating the radiation environment as applicable to health and welfare facilities and services;

-- Inspecting and estimating damages to hospital, medical, sanitary, welfare, and social security facilities, and food and drug stocks;

-- Locating food stocks and determining their fitness for human, animal or industrial use;

-- Recommending actions concerning the condemnation and embargo of contaminated foods, and the salvage and reprocessing of others.

-- Conducting epidemiological surveys and implementation of communicable disease control measures, including mass immunizations, obtaining vaccines, recommending sites for refuse disposal and for surveillance to prevent insect and rodent infestations, and recommending pesticides and how best to apply them;

-- Establishing mental health crisis counseling centers and obtaining official and professional agency personnel to operate these centers;

-- Reestablishing local health and welfare departments, Social Security Offices, and educational facilities, and restoring essential health services;

DEPARTMENT OF THE TREASURY is responsible for:

-- Directing any actions required to maintain or to reestablish the orderly operation of the financial system after a nuclear detonation or major dispersal of radioactive material, including (1) expediting the production and distribution of coin and currency to meet emergency demands; (2) expediting the processing of claims resulting from damage or destruction of currency; (3) providing a moratorium on calling funds on deposit with banks designated as tax and loan depositories; (4) permitting the pledging of government guarantees of loans for rehabilitation purposes as collateral for government deposits; (5) altering procedures pertaining to redemption or replacement of government securities; and (6) giving priority handling of claims for the loss or destruction of government checks;

OFFICE OF TELECOMMUNICATIONS POLICY is responsible for coordinating the development of plans and policies for the utilization of telecommunications resources in a peacetime nuclear emergency and shall be prepared to administer such telecommunications resources as may be required to cope with a peacetime nuclear emergency.

Within the DOJ, the FEDERAL BUREAU OF INVESTIGATION is responsible for:

-- Assisting DHEW with the identification of the dead.

For many years I've been doing investigative journalism, working for a large New York area daily newspaper and on radio and television. I also teach investigative reporting at the state university level. I've never come across an issue subjected to as extensive a cover up as nuclear power.

I became a journalist because of working for a newspaper which went by the philosophy: "Give light and the people will find their own way." In this book, I try to do that.

Some of the facts will be shocking. Some will be new. Some will be ugly. Much comes from the government and industry's own documents, information they have not wanted disclosed.

KARL GROSSMAN

Sag Harbor, N.Y.
1980

CHAPTER ONE

What Is At Stake?

Why is nuclear power dangerous?

Most of all, because of its radioactivity.
Webster's Dictionary provides this definition:

> **ra′di·o·ac·tiv′i·ty** (-ăk·tĭv′ĭ·tĭ), *n. Physics & Chem.* The property or process whereby certain elements or isotopes (notably radium, uranium, thorium and their products), whether free or combined, spontaneously emit particles and/or rays by the disintegration of the nuclei of their atoms. Cf. ALPHA RAY, BETA RAY, GAMMA RAYS. *Artificial radioactivity* may be induced by bombarding nuclei with particles, as from a cyclotron. — **ra′di·o·ac′tive** (-ăk′tĭv), **ra′di·o–ac′tive,** *adj.* —

Radioactivity destroys life—especially human life—by altering cells. A person will die within forty-eight hours of exposure to a very high dose of the ionizing radiation given off by radioactive elements.

Radioactivity throws a monkey wrench into nature; its ionizing radiation alters the electrical charge of cells—and the kickback is the end of life. Radioactivity is insidious because it cannot be seen or felt or smelled or heard or tasted; it comes in rays and sub-microscopic particles. It is invisible poison.

A dose of two or three thousand "rem" (a standard measure of radiation, an acronym for "Roentgen-equivalent-man") causes brain cells to swell and enlarge, the brain to press against the skull and hemorrhage; it causes fever, delirium, psychosis, loss of muscle control and, after a brief period of lucidity, death.

A dose of 500 rem will kill half the people exposed to it. Their cells cease dividing, their skin ulcerates, their hair falls out; they undergo vomiting, diarrhea and gastro-intestinal bleeding; their white blood cells and platelets (protection against infection and clotting factors) are destroyed: they die from infection or massive hemorrhage.

Survivors are likely to get cancer (most often, leukemia) and to suffer genetic damage and doom their still unborn children and their children's children to disease and deformity through mutations such as congenital heart disease, Mongoloidism, even Janus monstrosity (two faces on a single head and body).

Quick death or slower cancer and inheritable genetic damage are produced in direct proportion to the radiation dose, from high levels of 400, 300, 200, 100 rem right down to "low-level" radiation, ten rem to a millirem (a thousandth of a rem). Low doses allow an "incubation" or "latency" period of five to forty years before cancer or death strikes.

There is no safe level ofradiation exposure.

The effects of radioactivity are cumulative. A little from A-bomb fallout, a little from an X-ray, a little from the drift from Three Mile Island, a little from the "routine" (so-called low-level) radioactive discharges from a nuclear plant near your home, a little from the wind blowing from a uranium mine site many miles away, a little from the radioactive material you didn't know was in the storage hold on an airplane on which you were flying, a little from radioactive waste leaking into the water—a little then and a little now—every bit builds up in your body, and brings you closer to sickness and death.

Radioactivity in the environment is irreversible. It cannot be detoxified. Once let out by man, like the evils of Pandora's box, into the air, water or earth, it is here to stay, for the hazardous lifetime of these radioactive substances can run into thousands and millions and even billions of years.

Further: of all species, the human being is far more affected by radioactivity than many other forms of life. Cockroaches can absorb 200 times more radioactivity than humans before dying, some other insects still more.

And radioactive particles and rays attack the organs of the body selectively. Some attack the liver, some the thyroid gland, some the reproductive organs.

Human embryos, human infants, children are—in that order—most vulnerable to radioactivity, because their cells must grow rapidly. (That is why pregnant women and children were ordered evacuated from the area of Three Mile Island.)

In every step of nuclear power production, radioactivity is released into and contaminates the environment and the chain of life: when uranium is mined, in the extensive and expensive process when it is made into fuel, when it is shipped, when it is made to go through fission (or atom-splitting) in a reactor and, finally, when it ends up as waste, emitting radiation for thousands of times the extent of the earth's history.

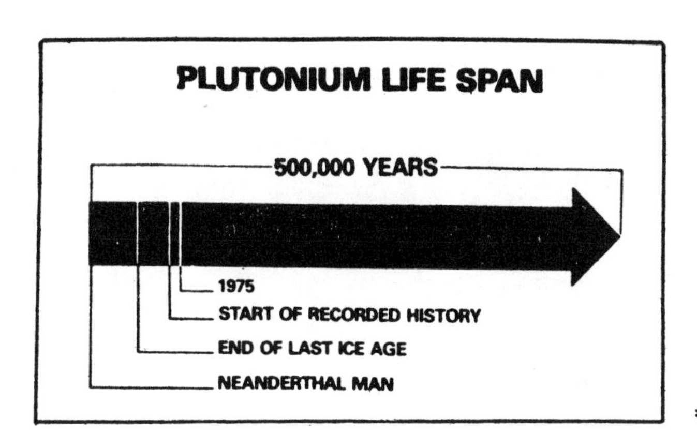

*

Radioactivity is routinely permitted to be discharged from the smokestacks of nuclear power plants and from pipes into the bodies of water alongside of which most nuclear plants are built to provide the massive amounts of coolant nuclear plants require. There is no possible way to hold all the radioactivity in. "Planned" releases are constant.

As radioactivity moves up the chain of nature its concentrations increase—like DDT. The ratio of concentration is designated "CF," for "concentration factor." Radioactivity in water is increased many thousand-fold in plants and plankton which draw from that water, and many thousand-fold again in fish which have eaten the plants and plankton, and then many thousand-fold again when you eat the fish. Radioactivity on pasture land is increased many thousand-fold when the grass becomes part of cow's milk, and again many thousand-fold when you drink the milk.

The U.S. government here illustrates how radiation routinely emitted from nuclear plants gets to man:

*Committee for Nuclear Responsibility, San Francisco, California, 94101.

APPENDIX H

EXAMPLES OF FIGURES SHOWING RADIATION EXPOSURE PATHWAYS

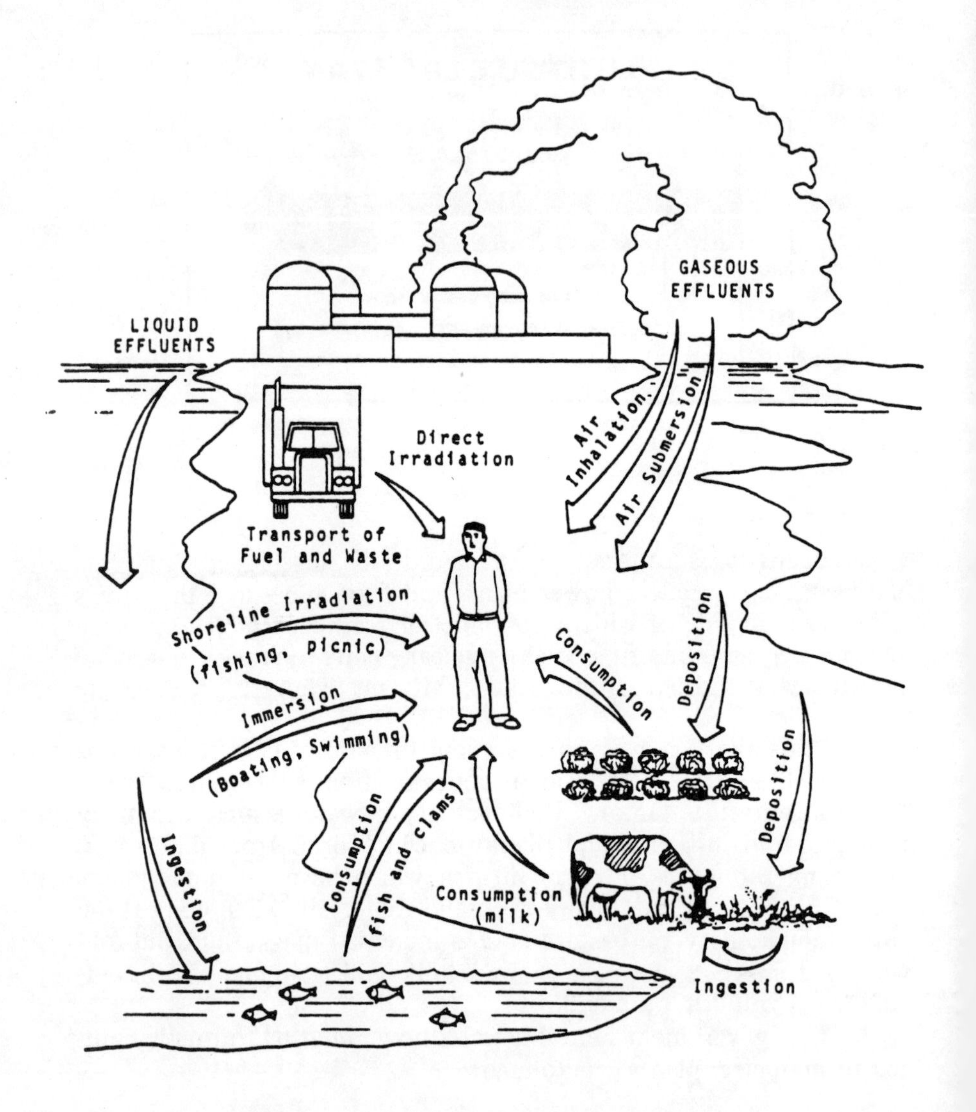

Figure H-1. Generalized Exposure Pathways for Man

From: U.S. Nuclear Regulatory Commission Regulatory Guide 4.2, Revision 2, "Preparation of Environmental Reports For Nuclear Power Stations," July, 1976.

Agencies promoting nuclear power have had to correct their estimates of expected casualties from it. A branch of the Nuclear Regulatory Commission, for instance, recently declared that it had miscalculated by a factor of 100,000—yes, underestimated by 100,000 times—the effects of the lethal radioactive gas called radon emitted from the mountainous piles of "mill tailings" amassed as uranium is mined for nuclear fuel.

This memorandum makes the point:

Appendix II

DOCKET NUMBER
PETITION RULE PRM-51-1

UNITED STATES
NUCLEAR REGULATORY COMMISSION
ATOMIC SAFETY AND LICENSING BOARD PANEL
WASHINGTON, D. C. 20555

September 21, 1977

MEMORANDUM FOR: James R. Yore, Chairman
Atomic Safety and Licensing Board Panel

FROM: Walter H. Jordan, ASLBP

SUBJECT: ERRORS IN 10 CFR §51.20, TABLE S-3

Since the radon continues to seep from the tailings pile for a very long time, the total dose to people over all future generations could become very large. Deaths in future generations due to cancer and genetic effects resulting from the radon from the uranium required to fuel a single reactor for one year can run into the hundreds. (See Pohl, Search, Vol. 7 No. 8, Aug. 1976). It is very difficult to argue that deaths to future generations are unimportant.

In summary the values given in Table S-3 for the amount of Rn-222 emitted per annual fuel requirement is grossly in error. So also is the dose to offsite population from milling due to one annual fuel requirement -- the correct number is more nearly 10 million person-rem rather than 100 person-rem.

The correct value would be some 100,000 times greater!

Far worse: with an accident at a nuclear plant, there is a potential for release of a gargantuan amount of radioactivity—more radioactivity than has ever been unleashed on earth. That is because a nuclear plant has many times more radioactive material contained in it than a nuclear weapon, typically 200,000 to 300,000 pounds of enriched uranium going through fission far longer than a weapon, thus producing vastly more radioactive poisons.

The basic text on nuclear plant accidents used by the nuclear industry and the U.S. government is called "THEORETICAL POSSIBILITIES AND CONSEQUENCES OF MAJOR ACCIDENTS IN LARGE NUCLEAR POWER PLANTS," (called "WASH–740" for the filing number of the U.S. Atomic Energy Commission which prepared it). It admits:

> It must be clearly recognized, however, that major releases of fission products from a nuclear power reactor conceivably could occur and that a serious threat to the health and safety of people over large areas could ensue.

"No one knows now or will ever know the exact magnitude" of the chances of a catastrophic nuclear plant accident happening, the document declares, but for a nuclear plant of 100 to 200 megawatts—a fifth to a tenth of the size of the nuclear plants being constructed today—and "in a characteristic power reactor location," WASH–740 makes these projections:

> For the three types of assumed accidents, the theoretical estimates indicated that personal damage might range from a lower limit of none injured or killed to an upper limit, in the worst case, of about 3400 killed and about 43,000 injured.
>
> Theoretical property damages ranged from a lower limit of about one half million dollars to an upper limit in the worst case of about seven billion dollars. This latter figure is largely due to assumed contamination of land with fission products.
>
> Under adverse combinations of the conditions considered, it was estimated that people could be killed at distances up to 15 miles, and injured at distances of about 45 miles. Land contamination could extend for greater distances.

WASH—740

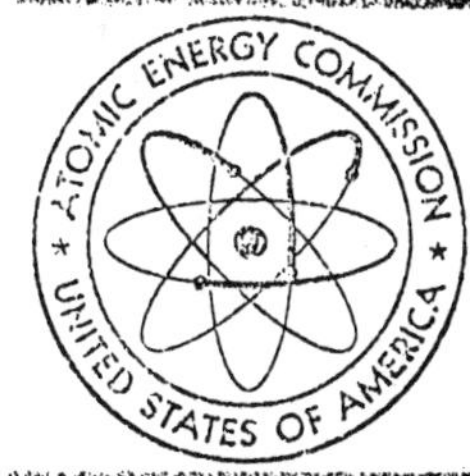

THEORETICAL POSSIBILITIES AND CONSEQUENCES OF MAJOR ACCIDENTS IN LARGE NUCLEAR POWER PLANTS

These 1957 casualty and damage estimates were substantially increased in a government analysis of the 1960's, the "WASH-740-update." That report was released only after a Freedom of Information Act challenge by Ralph Nader, Friends of the Earth and others.

Anticipated deaths from a single nuclear plant accident were increased by the U.S. government to 45,000, injuries to 100,000 and property damage to a range of $17 billion to $280 billion—what the government refers to repeatedly as "an appreciable fraction of the gross national product."

Note in this WASH-740-update document the suggestion in item 9 that "the results of the study must be revealed" to the Atomic Energy Commission and the Congressional Joint Committee on Atomic Energy (JCAE) "without subterfuge"; but telling the public is another matter. Item 10 suggests an increase in anticipated property damage for nuclear plant accidents "by a factor of 40"—to $280 billion per accident.

OFFICIAL USE ONLY

U. M. Staebler - 5 -

A meeting with the AIF Nuclear Safety Subcommittee will be arranged for late January or early February.

9. The results of the study must be revealed to the Commission and the JCAE without subterfuge although the method of presentation to the public has not been resolved at this time.
10. The results of the study suggest that the Price-Anderson liability level should not be reduced. Rather, an increase by a factor of 40 is suggested by the calculations (280 billion).
11. In view of the potential for damage from large reactor plants, a different attitude must be created on the type of safety programs undertaken, safety evaluations, and siting practices.

Enclosures:
1. Steering Committee attendees
2. Oultine - Topics on Probability

cc: CKBeck, REG
MBooth, DRD

On another page this extensive analysis, done more than a decade before the Three Mile Island near-catastrophe in Pennsylvania, declared:

> the possible size of the area of such a disaster might be equal to that of the State of Pennsylvania.

The statement was quoted in the film, "The China Syndrome." No, it did not originate in the head of a Hollywood screenwriter. It was taken from government deliberations.

Recent work by Dr. Richard Webb, nuclear engineer and world expert on nuclear plant accidents and their consequences, projects a million deaths from one catastrophic accident. Radioactivity would stream from a nuclear plant in a lethal seventy-five-mile long, one-mile wide cloud or plume which would travel with the wind. People caught in it would suffer acute radiation sickness and die. Over a vast area radioactivity would cause cancer and genetic damage. Tens of thousands of square miles of land might have to be abandoned permanently. Agriculture would be ruined for a century over an area the size of one half of the United States east of the Mississippi River (some 500,000 square miles).

Nor is that the worst. A major accident at a "breeder" reactor, which uses and produces plutonium, would not only disperse mammoth amounts of radioactivity–even more than a conventional nuclear plant–but would explode like a small atomic bomb, guaranteeing that its contents are released. Such breeder reactors have already been built in Europe and the Soviet Union; they are in the experimental stage in America. They are regarded by nuclear planners as the necessary "Phase II" for nuclear power because fissionable uranium is projected to run out in thirty years, long before oil, and man-made plutonium would have to be used to continue nuclear power.

How could a nuclear plant accident happen?

Very easily because of the very conditions of operation. Radioactive materials are used in a nuclear power plant just as coal, oil or gas are used in a conventional plant–as a heat source. Huge amounts of radioactive material are made to go through a chain reaction, a process in which atomic particles bombard the nuclei of atoms, causing them to break up and generate heat. The heat boils water, the steam turns a turbine and electricity is produced.

But to keep the nuclear reaction in check—to prevent the material from overheating—vast amounts of coolant are required, up to a million gallons of water a minute in the most common nuclear plants being built today ("light water" reactors). That is why nuclear plants are sited along rivers and bays, to use the water as coolant. If the water which cools the reactor "core"—its 200,000 to 300,000 pounds of radioactive fuel load—stops flowing completely (and a pipe rupture or a break in the reactor "vessel" could do that) there is only *fifteen to thirty seconds* for the "emergency core cooling system" (ECCS) to send water in. This is called a loss-of-coolant accident (LOCA).

If that emergency system, which has failed in many tests, fails to operate within fifteen to thirty seconds an unstoppable "meltdown" results:

> Catastrophic Nuclear Reactor Accidents 93
>
> 4.10 Consequences of an ECCS Failure
>
> It is in the critical time period from about 15 seconds after the rupture to 30 seconds that control of the accident must be gained by ECCS operation and the fuel temperature excursions halted. The greatest importance is attached to the need for having an adequate cooling water flow upward through the core in this period. If the vertical flooding rate, once emergency coolant reaches the core bottom, is below some critical value, presently believed to be in the vicinity of 0.7 inches per second, then the accident will proceed out of control.*

The core of nuclear fuel, now 5,000 degrees Fahrenheit, burns through the cement bottom of the nuclear plant and bores through

*From: *The Nuclear Fuel Cycle,* prepared by Union of Concerned Scientists, Massachusetts Institute of Technology Press, Cambridge, Mass., 1975.

the earth. This is what nuclear scientists have dubbed the "China syndrome." The white-hot core doesn't go to China, however, but to the water table underlying the plant. Then, in a violent reaction, molten core and cold water combine, creating steam explosions which can breach the containment and release a thousand times more radioactivity than the Hiroshima A-bomb.*

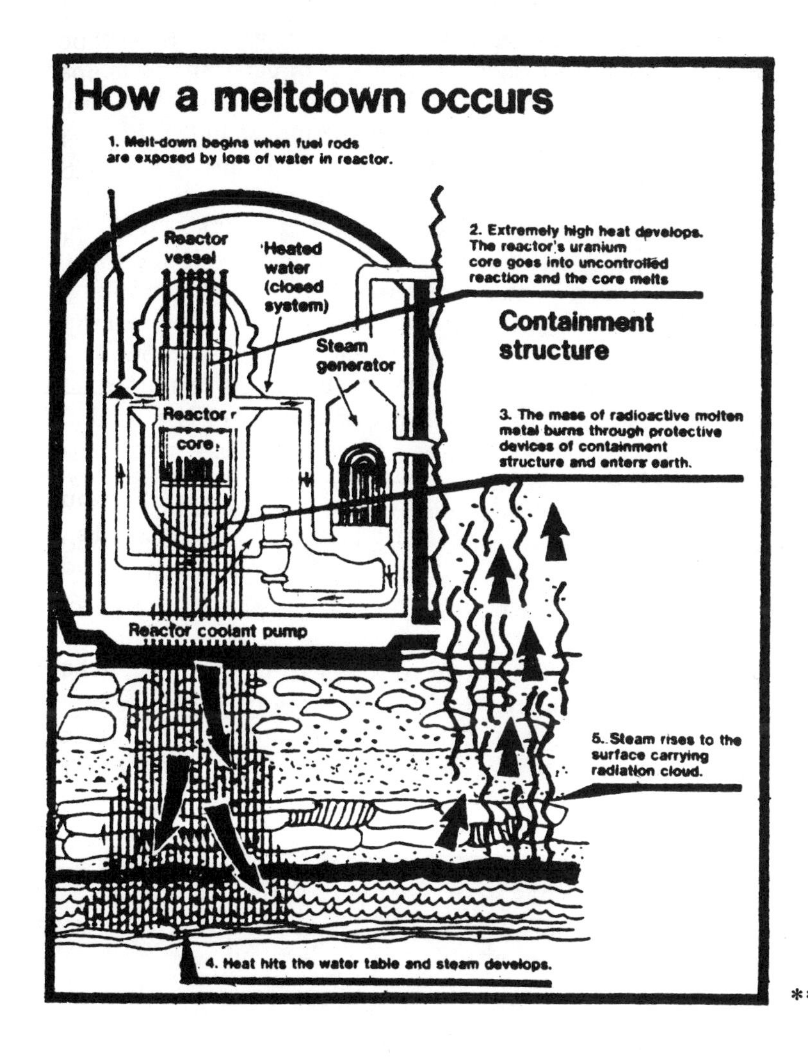

**

*Which contained only 20 pounds of radioactive material, more enriched.
**Citizens Energy Council, Allendale, N.J. 07401.

There are other kinds of nuclear plant accidents–including a "nuclear runaway" accident, which is even worse than a meltdown because an instant explosion occurs. We will provide details on all types of accidents, those that can happen and those that have already happened–although the public has, for the most part, not been informed.

And that purportedly mysterious "hydrogen bubble" at Three Mile Island was no isolated incident.

Eruption of hydrogen gas as a first reaction in a partial loss of coolant accident has been discussed fearfully in government and nuclear industry literature for decades. That is because a highly volatile substance called zirconium was chosen as the major material for the rods into which the radioactive fuel is loaded. There are 30,000 to 40,000 rods composed of twenty tons of zirconium in an average reactor. Many other substances were tried first, particularly stainless steel, but only zirconium worked well. Zirconium is used because it allows neutrons from the fuel pellets in the rods to pass freely between the rods and so to sustain a chain reaction. But zirconium has the great disadvantage that it is highly volatile and, when hot, will explode spontaneously upon contact with air, water or steam. Zirconium dust in air will explode instantly, for which reason a speck of it is used in flashbulbs, its main use. It will not erupt spontaneously in solid form except with heat. But it is heat that builds up, a great deal of heat in a very short time, with any interruption of coolant flow in a nuclear reactor. Zirconium, with the explosive power, pound for pound, of nitroglycerine, catches fire and has the potential to explode at a temperature of 2,000° F–well below the 5,000 degree temperature of a meltdown.

Before that, zirconium reacts to the heat by drawing oxygen from water and steam and letting off hydrogen, which itself can explode as well as further block coolant; and in its formation even more heat is developed, bringing the zirconium closer and closer to its explosive level.

The hydrogen bubble at Three Mile Island was no mystery but precisely what was and is expected in a partial loss of coolant accident. The cause for worry has been what is described in nuclear accident studies as a zirconium "metal-water" reaction, a reaction which can easily trigger a meltdown or full-scale nuclear runaway.

What is plutonium? How does it come into the picture?

Plutonium is the most potent radioactive substance in a nuclear plant, in any phase of the nuclear cycle. It is an element developed at the dawn of World War II for use in atomic bombs.

It is the most toxic substance in the universe.

It has the appropriate name of Pluto, the God of Hell. One of its discoverers described it as having "fiendishly toxic" power. It does not normally occur in nature but is a by-product of uranium fission.

A pound of plutonium, released as airborne dust, has the potential to cause fatal lung cancer in nine billion people—over twice the population of the earth today. An ounce (only a tablespoonful, because plutonium is very heavy) can kill 200 million people. A millionth of a gram of plutonium will cause cancer.

An average nuclear power plant produces 400 to 500 pounds of plutonium yearly as a waste product. The amount of plutonium in a plant's nuclear core at any time during its operation depends on the "age" of the core; it can be up to 1,000 pounds of plutonium. The amount varies because, periodically, reactors are "refueled"—older fuel rods, heavy with plutonium and other radioactive wastes are taken out and new rods put in.

Plutonium has power of its own. The strategy of those in the nuclear power business is to use this plutonium waste, as high grade uranium runs out in coming years, for fuel in nuclear plants.

"Breeder" reactors are being built which not only utilize plutonium as a fuel but produce even more plutonium waste than conventional reactors—"breed" the stuff—which can then be re-used as fuel. This kind of perpetual motion theory using plutonium is called by those in nuclear power "the plutonium economy"—endless energy EXCEPT, and they are some exceptions:

- A large breeder reactor contains seven tons of plutonium which, dispersed in a single catastrophic accident, would spread enough of the most toxic radioactive substance in the cosmos to kill every person in the world 42,000 times over!

- This super-concentrated radioactive poison stays in the environment for 500,000 years.

- Plutonium breeders are even more unstable than regular nuclear power plants. *We Almost Lost Detroit** is the true story of what almost happened when the Fermi breeder near Detroit underwent an accident in 1966; the city narrowly avoiding catastrophe.

- In a breeder, liquid sodium is used as coolant, not water. Liquid sodium will explode and burn on contact with air or water. A pinhole liquid sodium leak in a breeder can lead to rapid disaster.

*John G. Fuller, Readers Digest Press, New York, 1975.

• A breeder can explode like a small atomic bomb, shattering its concrete containment as if it were an egg shell and unleashing all radioactive materials inside, including the vast tonnage of lethal plutonium.

"We nuclear people have made a Faustian bargain with society," said one leading plutonium breeder advocate, Alvin Weinberg, in 1972.* Dr. Weinberg, former director of Oak Ridge National Laboratory, one of several U.S. national laboratories which serve as taxpayer-supported research and development facilities for the nuclear industry, speaks of a "nuclear priesthood" in charge.

With the plutonium breeder, one single accident spells doomsday and "the issues involved are primarily not scientific or technological, but moral and ethical," as the U.S. National Council of Churches declared in its 1976 "Resolution on 'The Plutonium Economy'." "We are talking literally about the future of humanity," said the Council. "We are charged by God to be caretakers of Creation," it declared, calling for a stop "in the continuing development of plutonium use, during which society can determine its options and decide responsibly on the avenue it wishes to follow, before irreversible commitments have been made."

If nuclear plants are so dangerous, why the push to build them?

It began with the scramble at the end of World War II among the nuclear scientists in the U.S. who had worked making atomic bombs in the "Manhattan Project." The war was over and, although many were able to keep on making atomic weapons, what about the rest? (Two decades later the aerospace engineers of America faced a similar threat of unemployment when the U.S. cooled on rocket ships.)

These nuclear men joined forces with their bureaucratic governmental counterparts and in 1946 created the Atomic Energy Commission, which slickly—with a lot of public relations media-hoodwinking developed as a matter of wartime censorship in the Manhattan Project—pushed a scheme called "atoms for peace."

The U.S. military encouraged and supported the effort because it saw that wide activity in the field served its own interests: plutonium made for power plants could be used for bombs, too, and a commercial nuclear program would help justify and shield huge military nuclear expenditures. It could be a joint operation.

Also, in America production of nuclear weaponry was principally

*"Social Institutions and Nuclear Energy," *Science*, July 7, 1972.

contracted out by the government to large corporations which wanted to see their nuclear business now expand in every way possible.

But the electric utilities, traditionally conservative institutions, wanted nothing of it. They pointed to their liability in the event of nuclear accidents and demanded: who would pay for damage to property, for injury and death?

Insurance companies wouldn't accept responsibility. "The catastrophe hazard is apparently many times as great as anything previously known in industry," Hubert W. Yount, then vice-president of Liberty Mutual Insurance testified before the Congressional Joint Committee on Atomic Energy. "We have heard estimates of catastrophe potential under the worst possible circumstances running not merely into millions or tens of millions but into hundreds of millions and billions of dollars. It is a reasonable question of public policy as to whether a hazard of this magnitude should be permitted . . . . Obviously there is no principle of insurance which can be applied to a single location where the potential loss approached such astronomical proportions. Even if insurance could be found, there is a serious question whether the amount of damage to persons and property would be worth the possible benefit accruing from atomic development."

The nuclear establishment of government scientists and bureaucrats then threatened the utilities. If you don't build nuclear power plants, they said, the government will.

And indeed, at Shippingport, Pennsylvania, Admiral Hyman Rickover–with government funding and manpower–began building the nation's first civilian nuclear reactor. Government literature on nuclear power began referring to "government controlled sites" for electric generation.

If we build them, the intimidated utilities then asked, what about the legal liability if and when a nuclear plant erupts? to cover this, the nuclear lobby arranged for the U.S. Congress–which remains malleable to its desires–to pass the Price-Anderson Act. It limited liability for a nuclear plant accident, no matter how many people are killed and injured and how many billions are lost, to $560 million–with the government picking up the first $500 million. That's why there is the "nuclear clause" on insurance policies in America.

Here is the Price-Anderson Act, which became Section e of the Atomic Energy Act.

PART I. THE ATOMIC ENERGY ACT

Public Law 83–703
(68 Stat. 919)

"CHAPTER 1. DECLARATION, FINDINGS, AND PURPOSE

"SECTION 1. DECLARATION.—Atomic energy is capable of application for peaceful as well as military purposes. It is therefore declared to be the policy of the United States that—

"a. the development, use, and control of atomic energy shall be directed so as to make the maximum contribution to the general welfare, subject at all times to the paramount objective of making the maximum contribution to the common defense and security; and

---

Aggregate liabilities.

"e. The aggregate liability for a single nuclear incident of persons indemnified, including the reasonable costs of investigating and settling claims and defending suits for damage, shall not exceed (1) the sum of $500,-000,000 together with the amount of financial protection required of the licensee or contractor or (2) if the amount of financial protection required of the licensee exceeds $60,000,000, such aggregate liability shall not exceed the sum of $560,000,000

With their liability all but eliminated and other subsidies offered, the utilities found the nuclear business to their liking after all. The rate of profit of utilities through the U.S. is set on the basis of capital expenditure. The more money a utility spends on construction, the more money regulatory agencies let it make. Each new billion dollar nuclear plant constructed by a utility, even when ratepayers' money is used, allows it to make many millions of dollars more. The utilities also enjoyed the benefits of having nuclear research and development done for them for free at government laboratories at a cost to taxpayers of billions of dollars since World War II.

To top it off, the multinational oil giants began dominating the nuclear field, particularly the Standard Oil Trust, led by Exxon, as well as Gulf, Kerr-McGee, Continental and Getty (notice the little atom within the "G" of a Getty gas station sign).

These oil companies went on to buy up most of the uranium mines and reserves as well as milling and fuel fabrication centers, becoming, in short order, principals or being interlocked in all phases of the nuclear "cycle."

In the 1950's, as this was happening, a U.S Presidential Commission led by former CBS chairman William Paley issued a report called "Resources for Freedom" stressing solar power development. "It is time for aggressive research into the whole field of solar energy—an effort in which the United States could make an immense contribution to the welfare of the free world," said the commission.

But that never happened because solar power and all the forms which could be produced right at one's home and place of work (people having their own power supply in a decentralized way using infinite or renewable, ever-available energy) constituted a profound threat to the existing energy system and those in charge of it. If resources would be used to build and install the hardware so that homes and businesses could tap the sun and the wind and be energy-independent, there would be no need to pay for fuel after that or to be plugged into a centralized system of power.

The power brokers preferred to keep their centralized system intact. With conversion to nuclear power only the heat source need be changed. So they moved to lock the U.S. into nuclear power.

And America's nuclear establishment pushed—in some instances all but gave away—its product worldwide so that the U.S light water reactor became the dominant nuclear power technology. This exportation increased in the 1970's as anxiety over nuclear power in the U.S. led the industry and government, in order to keep the industry afloat, to stress markets in developing countries. It has been a grand attempt to hook the world.

But isn't nuclear power a cheap energy source?

Repeated studies show just the opposite, despite this being a favorite industry line.

A recent House of Representatives' report states that it is very costly indeed.

95TH CONGRESS } HOUSE OF REPRESENTATIVES { REPORT
2d Session { No. 95–1090

NUCLEAR POWER COSTS

APRIL 26, 1978.—Committed to the Committee of the Whole House on the State of the Union and ordered to be printed

BASED ON A STUDY BY THE ENVIRONMENT, ENERGY, AND NATURAL RESOURCES SUBCOMMITTEE

On April 12, 1978, the Committee on Government Operations approved and adopted a report entitled "Nuclear Power Costs." The chairman was directed to transmit a copy to the Speaker of the House.

I. ABSTRACT

Contrary to widespread belief, nuclear power is no longer a cheap energy source. In fact, when the still unknown costs of radioactive waste and spent nuclear fuel management, decommissioning and perpetual care are finally included in the rate base, nuclear power may prove to be much more expensive than conventional energy sources such as coal, and may well not be economically competitive with safe, renewable resource energy alternatives such as solar power. Nuclear power is the only energy technology which has a major capitalization cost at the outset of the fuel cycle and at the end of the fuel cycle. As the cost of nuclear energy continues to climb, and as a solution to the problems of radioactive waste management continues to elude government and industry, States such as California are rejecting the increased use of nuclear power and favoring the greater use of renewable energy technologies. These developments and others discussed in this report raise major questions for Federal decisionmakers about how best to cope with the Nation's energy crisis in the years ahead. Practical recommendations aimed at greater economy, efficiency, and effectiveness in government actions are proposed.

The same Congressional report stressed: "If the federal government spent only a small portion of what it has already spent on nuclear power development for the commercialization of solar power, solar generated electricity would be economically feasible within five years."

After analyzing the rates of nuclear and non-nuclear utilities in America, Ralph Nader's Public Citizens' Critical Mass Energy Project and the Environmental Action Foundation declared "the use of nuclear power to generate electricity has usually resulted in higher utility rates for consumers" and the "long-held claims of consumer savings is largely a myth. There is a strong correlation between the use of nuclear power and the rise in electricity rates."

And things will get worse as uranium becomes more scarce.

What about jobs?

Nuclear power provides very few jobs. A nuclear plant is manned by only 100 workers. "Nuclear power plants are capital intensive and thus produce few jobs," as the Congressional "Nuclear Power Costs" report puts it, emphasizing that the "renewable energy sources such as solar and conservation" produce many jobs.

A major study by the Council on Economic Priorities projects more than twice as many jobs created for the energy produced if the same amount of dollars are invested in solar heating and cooling, energy-saving re-insulation and some thirty-four available measures. And this takes into full account construction jobs in building a nuclear facility. These jobs quickly vanish, while the alternative options create jobs for solar installers and maintenance people, plumbers, sheetmetal workers, home improvement workers and many others for a long time to come.

CHAPTER TWO

How It Works

What is the principle behind nuclear power?

Basically it's just an elaborate—and the world's most dangerous—way of boiling water. Nor is it so complicated or so involved that only a "priesthood" of nuclear scientists and assorted experts can understand it. You easily can, too.

No matter what the type of nuclear plant, all any of them do is boil water. Steam is given off and the steam turns a turbine which produces electricity.

In the First Century A.D., the Greeks worked with the idea that if water was boiled in an enclosed space, steam would form and high pressure would build up which could produce motion.

In the 1700's,the steam engine first came into use. In England and then in America and Europe, coal and wood were used to fire up these devices—which, ever since they have been around, have blown up on occasion, a result of the pressure under which they function.

In the 1800's came electricity. Take a conductor like a coil of copper wire or a copper wheel, spin it between the poles of a magnet, and it will generate electricity which can then be transported through wires.

But what could keep that wheel turning?

Water movement can do it, from hydroelectric units along streams and rivers to those catching the cascading water at Niagara Falls. The wind can do it, as the windmills which studded much of the world for centuries testify. An internal combustion engine, such as an auto engine turning a generator in a car, can do it.

There are also ways to produce electricity without the need for energy spinning a wheel. For example, a battery generates electricity through chemical action. And solar or photovoltaic cells—the main sources of power in space satellites—generate electricity by using the sunlight which shines on them.

But by the mid-twentieth century small, decentralized energy operations using water and wind power had fallen under the weight of giant utility monopolies making steam-generated electricity in big, centralized power plants. They used coal or oil or gas to boil water, to produce steam, to turn a turbine, to make electricity.

By the 1950's, a crossroads—which we really are still at—was reached. There could have been and can be a renewal of decentralized energy production, making use of the best of modern technology to tap an array of abundant natural energy sources.

2.
1.
REG.OPERATIO
FULL STEAM
SCRAM
DECONTAMINATE
1
2
3
CLASSES OF
EXCURSION
DECOMMISSIO
98765
GE
3.
SIMPLIFIED DIAGRAM OF NUCLEAR POWER STATION, SHOWING THE HEAT OF FISSION IN REACTOR FUEL ASSEMBLY (1) HEATING WATER TO PRODUCE THE PRESSURE OF STEAM (2). THIS IS WHAT TURNS THE TURBINES (NOT SHOWN) TO PRODUCE ELECTRICITY (3)...

Vań Howell

The notion of nuclear power was grafted onto the existing system of having big, centralized power plants boil water, but instead of producing heat through the combustion of coal or oil or gas, a nuclear plant produces heat through breaking up what is often called the building block of the universe—the atom—through fission.

What is an atom? What is fission?

In Greek, atom means indivisible and chemically atoms cannot be broken up. All elements are made up of atoms. Although it would take twenty-five million atoms to cover the head of a pin, each atom is like a miniature solar system with a nucleus made up of protons and neutrons; around the nucleus electrons spin like planets around the sun.

Holding the nucleus together is what is called binding energy. Certain complex atoms having many neutrons and protons in their nuclei are not firmly bound.

They fall apart, break up—are radioactive. Their nuclei disintegrate and they send out particles and rays. But those radioactive elements found in nature, for example radium and uranium, are only feebly radioactive. That is because, like all naturally radioactive elements, they originated in the formation of the earth and have been disintegrating ever since. Eventually—in still more millions or billions of years—they will lose all their radioactivity and settle into stable elements.

Uranium-235, for instance, which has a total of 235 protons and neutrons (thus the number 235) takes 710 million years before just one half of it breaks down and becomes what uranium settles into naturally, lead. (The time required for one half of a radioactive substance to break down is called its "half-life." Twenty times that figure is the time it would take to break down completely and transform into a stable element. In the case of uranium this is 710 million years times twenty or 14.2 billion years.)

In the 1930's, after neutrons were first detected, scientists were experimenting by bombarding various elements with neutrons "to see what would happen," as the U.S. government recounted in its 1963 history of nuclear power, *Our Atomic World.* The source of the bombarding neutrons were elements found to emit them.

Of major interest was uranium. Experiments were conducted to see if neutron bombardment could alter its precarious yet slowly-changing atomic balance suddenly, could make what was an already unstable element even more unstable, could cause it to shatter into new chemical elements.

In the late 1930's in Germany, fission—the splitting or fracturing of an atom—was first achieved. Uranium-235 came apart under neutron bombardment.

But it wasn't as simple as splitting a nucleus into two neat, separate pieces. As work on fission proceeded, it became clear that when fission occurred, uranium-235 split into two (sometimes three) principal particles, but these were in 200 varieties (or isotopes) of mostly highly unstable, intensely radioactive twins of safe, stable elements in nature. Among these were radioactive iodine, strontium-90, radioactive cesium, all called *fission products.* And great amounts of radioactivity were produced in the process of fission, in all three forms of ionizing radiation: radioactive rays, called "gamma" rays (nearly identical to X-rays), "alpha" particles (made up of two protons and two neutrons) and "beta" particles (electrons). There was heat and, importantly, additional neutrons.

Van Howell

Breaking the cosmic building block came at huge potential cost to life. The ionizing radiation which fission produces can disrupt the electrical balance of living cells, causing death, cancer and genetic damage. Gamma rays can penetrate three feet of cement. Beta particles can penetrate human skin; only metal can stop them. Both alpha and beta particles, inhaled in air or swallowed in food, move on to destroy body cells. Unleashed neutrons themselves can swiftly kill, as in the neutron bomb. In unlocking the atom we allow these poisons to come rushing out in great quantity.

But in 1939 war was sweeping the world, and it was feared by scientists in the U.S., particularly refugees from the Nazis, that Germany might be turning fission into a war weapon—a bomb—by using the potential "chain reaction" of fissionable material.

If enough uranium could be assembled in one place and fission begun, it was theorized, the additional neutrons (erupting with fission products and heat and radioactivity) would crash into other uranium nuclei, releasing more neutrons along with more fission products, heat and radioactivity. This process would continue in a "chain reaction."

Albert Einstein wrote that year to President Franklin D. Roosevelt saying "that it may have become possible to set up a nuclear chain reaction in a large mass of uranium by which vast amounts of power and large quantities of new radium-like elements would be generated," that "extremely powerful bombs of a new type may thus be constructed. A single bomb of this type, carried by boat and exploded in a port, might well destroy the whole port together with some of the surrounding territory."

In December 1942, in Chicago, the first chain reaction succeeded—a pivotal step towards nuclear bombs and nuclear power, two sides of the same coin.

From the government's *Our Atomic World:*

The Fission Bomb Is Exploded

The American scientists present on that historic December day were part of the tremendous super-secret scientific and industrial complex that bore the unrevealing title Manhattan District. The United States had been at war almost a year. An uncontrolled fission reaction gave promise of producing an explosion of untold proportions. This promise, coupled with the possibility that enemy scientists might be nearing such a goal, had launched a vast Allied effort.

The Manhattan Project, as it was commonly known, included a variety of "hush-hush" facilities. Each of these installations, in New York, Illinois, Tennessee, New Mexico, California, and Washington, had its own experts working night and day to solve the baffling problems surrounding development of a fission weapon.

A key problem was that only uranium-235, which comprises a tiny fraction (a mere 0.7 per cent) of the uranium found in nature is fissionable—can be split by neutron bombardment. Uranium ore consists mostly of non-fissionable uranium-238.

So the government proceeded in two ways. On one hand it began attempting to separate uranium-235 from uranium-238.

Several methods of achieving large-scale separation were tried. The most successful and economical, known as "gaseous diffusion," involves compressing normal uranium, in the form of uranium hexafluoride gas, against a porous barrier containing millions of holes, each smaller than two-millionths of an inch. Since the ^{235}U molecules are slightly lighter than the ^{238}U, they bounce against the barrier more frequently and have a greater chance of penetrating. Thus, although the gas at first contains only 0.7% ^{235}U, the process of compression is repeated several thousand times, and the proportion gradually increases until the necessary concentration is reached.

For this operation an enormous plant containing a very large barrier area, miles of piping, and countless pumps was built at Oak Ridge, Tennessee.

On the other, it proceeded to make an atomic bomb out of a man-made substance found to be fissionable—plutonium.

Plutonium was created in 1941 by four U.S. scientists*working with uranium-238. They found that, though U-238 did not split under neutron bombardment, it absorbed a neutron and became a new element. They named it plutonium-239.

It was "as fissionable as uranium-235 and hence theoretically just as feasible for a bomb," says *Our Atomic World,* which goes on:

*Glenn Seaborg, Joseph Kennedy, Arthur Wahl, Emilio Segre.

The way to manufacture usable amounts of plutonium, an element that had never before been detected on earth, is to expose uranium to a very intense neutron bombardment. The best-known place to find a rich supply of neutrons was the heart of a self-sustaining chain-reacting pile of uranium. Accordingly, very large piles, or *reactors*, were rushed to completion near the Columbia River at Hanford, Washington, to make plutonium.

First atomic bomb explosion at Alamagordo, New Mexico, at 5:30 a.m. on July 16, 1945.
Courtesy U. S. Army

On July 16, 1945, a plutonium bomb, carefully assembled by another group of scientists at "Project Y," Los Alamos, New Mexico, was successfully tested in the New Mexico desert. The heat from that first man-made nuclear explosion completely vaporized a tall steel tower and melted several acres of surrounding surface sand. The flash of light was the brightest the earth had ever witnessed.

A ^{235}U bomb was dropped on Hiroshima, Japan, on August 6, 1945. Three days later a plutonium bomb was dropped on Nagasaki, Japan. Hostilities ended on August 14, 1945.

Nuclear Energy Is Needed for the Future

The chief source of the enormous quantities of energy used daily by modern civilization is fossil fuels in the form of coal, petroleum, and natural gas. Concentrated sources of these fuels, though large, are far from inexhaustible, and it has been said that future historians may refer to the brief time when they were used as "the fossil-fuel incident."

The next great source of energy will probably be nuclear reactors, in which controlled chain reactions release energy from the large store of fissionable materials in the world.

The above, the jump from the dropping of atomic bombs to the claim that "nuclear energy is needed" is how the government's nuclear history continues on—both in the book and in reality.

In 1945, at the end of World War II, there was a huge vested interest in what the government concedes was a "scientific and industrial complex" building nuclear bombs under its aegis. Hundreds of thousands of people were involved. A network of government-built facilities was contracted out to major corporations to run. They were bent on commercializing this deadly-dangerous undertaking of atom-splitting, to keep the business expanding, and to perpetuate it. Two basic reactor designs were subsequently developed by the U.S. and now are the dominant nuclear plant designs throughout the world. They are the boiling water and pressurized water reactors.

How do nuclear power plants work?

Like conventional power plants: water is boiled, steam turns a turbine and electricity is produced. The difference is only in the heat source—the fuel used to boil the water.

In a nuclear plant the heat comes through the fission chain reaction, the same process as in an atomic bomb. In a boiling water or pressurized water reactor the uranium fuel is not supposed to be concentrated enough to allow the chain reaction to go out of control as it does in an A-bomb which contains highly enriched or concentrated uranium-235. This, however, is in dispute. In an A-bomb ten to twenty pounds of uranium enriched to a ninety per cent concentration of uranium-235 is brought together suddenly. A "critical mass" is formed, fission occurs and simultaneously an "implosion mechanism"—explosives—increase the density of the uranium. BAM!!! There is a blast, heat at a million degrees Fahrenheit, and fission products are released as radioactive fallout—a comparatively small amount, some two pounds, because of the speed of it all—but enough to devastate a population.

In a nuclear plant there is far more uranium: 200,000 to 300,000 pounds, enriched to three per cent uranium-235 (much from the same U.S. government-owned Union Carbide-run plant at Oak Ridge, Tennessee where weapons-grade fuel has been prepared since the war).

This vast amount of uranium goes through fission not for a fraction of a second as in a nuclear bomb but for years. And once it begins operation, fission products build up, with two tons eventually accumulating in an average plant.

Nuclear plant fuel is made in the form of half-inch long pellets which are packed into very thin twelve to fourteen-foot long "fuel rods" made of an alloy of zirconium, a highly volatile metal but regarded as the best "cladding" for nuclear fuel because it allows neutron flow.

The rods are put together in a "fuel assembly," generally 100 rods to an assembly, with spaces between the rods for water to circulate. There are hundreds of assemblies in a nuclear plant "fuel load" holding approximately 40,000 rods.

A Fuel Assembly

WASH-1250, U.S. Atomic Energy Commission

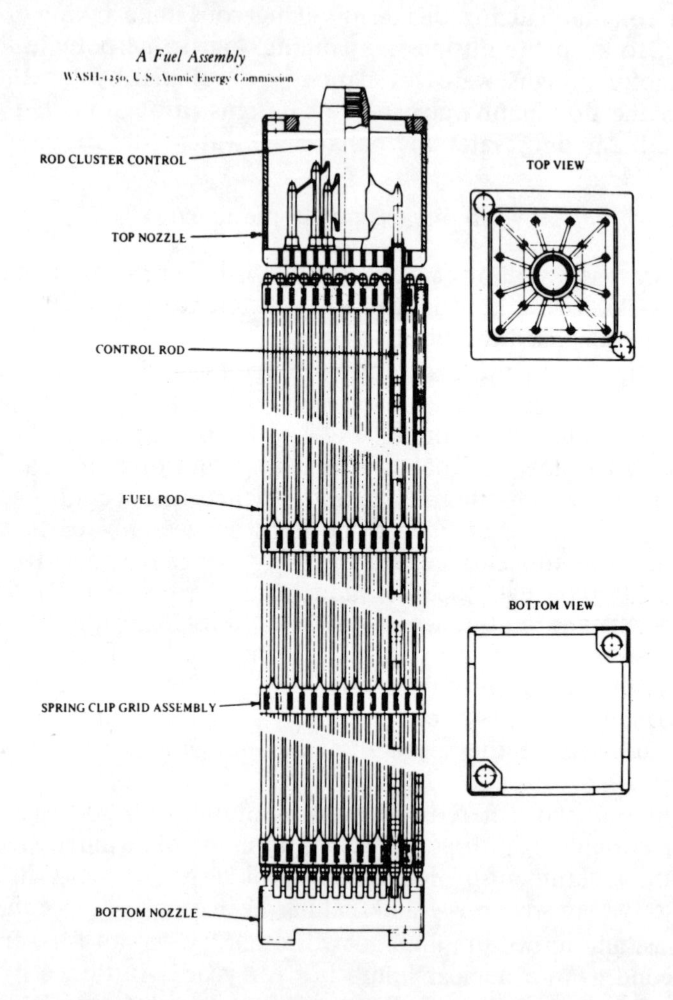

This load all goes into a bullet-shaped steel chamber forty to fifty feet high, fifteen to twenty feet wide. This is the reactor pressure "vessel." One fuel assembly is placed in the vessel at a time. The geometric arrangement of the fuel rods permits fission to begin. Fission is controlled by "control rods" made of boron or cadmium, two elements which absorb neutrons. As many as 177 control rods are placed at intervals amid the assemblies. During fueling the control rods are fully inserted. This serves in nuclear jargon to "poison" or quiet any fission by absorbing neutrons so there are not enough available to sustain a chain reaction.

When a reactor is ready to start up or "go critical" the control rods are withdrawn slowly; neutrons fly and fission begins. One act of fission creates others in a chain reaction which produces fission products, heat and radioactivity—and plutonium (from uranium-238 which comprises the bulk of the fuel load). These fission products and plutonium build up as the uranium-235 is depleted: every year most reactors are "re-fueled," a third of their fuel rods taken out and replaced with new fuel rods. When a reactor is in operation, the chain reaction is not constant. The control rods must be carefully pushed in and pulled out to regulate the rate of fission.

In the boiling water or pressurized water reactors, water works to both keep the fuel from overheating and to slow down or "moderate" neutrons to sustain the chain reaction.

In boiling water reactors, all of which are made by the General Electric Corporation, the nuclear fuel boils the water circulating around the fuel rods—which is at 1,000 pounds per square inch pressure—and the steam from this boiling water turns the turbine. Here is a diagram of a boiling water reactor:

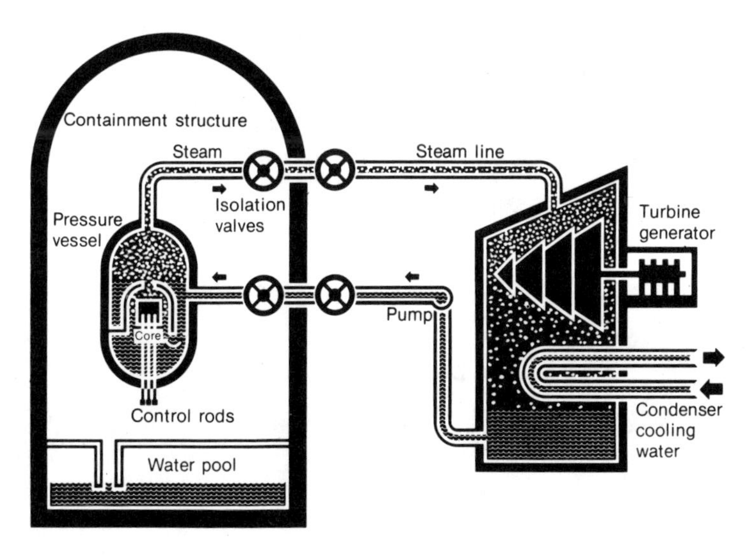

Pressurized water reactors, primarily made by the Westinghouse Corporation, have two water loops. In the first, water is circulated around the fuel at even higher pressure than in the boiling water reactor, some 2,250 pounds per square inch. (Keeping water under pressure prevents it from boiling at the normal 212 degrees Fahrenheit. In a boiling water reactor, water boils at 545 degrees; in a pressurized water reactor, the pressure is so strong that the water in the first loop doesn't boil, though it is heated to 600 degrees.) This searingly hot water in the first loop of a pressurized water reactor goes to thousands of thin tubes to heat another separate loop of water, which boils; its steam turns the turbine. Here's a diagram of a pressurized water reactor:

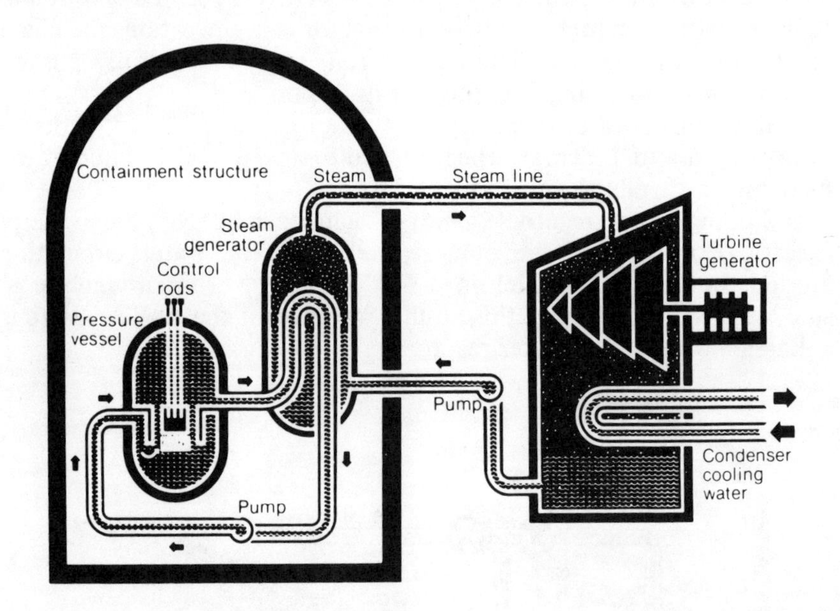

What is a breeder reactor?

In what is called a "breeder" reactor, plutonium is used as the fuel. Breeders are regarded as necessary if nuclear power is to continue, because uranium-235 is projected to run out within thirty years, despite the claim of *Our Atomic World* that there is a "large store of fissionable materials in the world."

When oil is still available, uranium-235, the basic fissionable material, will be gone. So, like the scientists on the Manhattan Project who resorted to plutonium for the Nagasaki bomb when sufficient uranium-235 could not be obtained, the nuclear establishment plans to resort to man-made plutonium in order to continue using nuclear power. It would mainly be made in breeders.

A breeder uses plutonium-239 surrounded by a "blanket" of uranium-238. Liquid sodium (instead of water) is the coolant because it can transfer heat well and does not slow down or moderate neutron flow (unlike water). In a breeder the neutron action must be fast to keep neutrons flying into the uranium-238 and "breeding" plutonium-239 out of it.

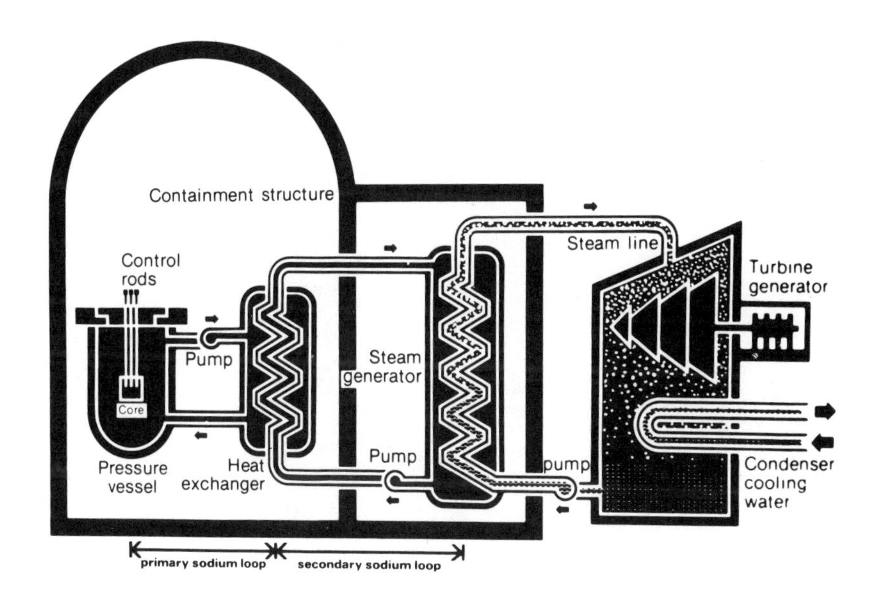

Sodium, however, reacts violently on contact with water or air. In a first loop, molten sodium (it is a liquid metal, hence the breeder's formal name, "liquid metal fast breeder reactor") circulates around the fuel. This sodium becomes highly radioactive. Fearing that a fire or explosion would send radioactive sodium into the environment if this loop came into contact with water, designers included a second loop also containing sodium. Heated by the first loop, it in turn boils the water in a third loop, producing the steam which turns a turbine.

The hypothesis has been that at least the sodium in the second loop would not be radioactive. Then, if sodium interacted with the water in the final loop and created a fire or explosion, the result should be less serious.

This hypothesis is highly questionable. Breeder reactors are regarded as even more fraught with danger than the standard boiling water or pressurized water nuclear power plants of today. Concentrations of plutonium in a breeder are so high that, as even the government and the nuclear industry admit, an accident in a breeder can cause it to explode like a small atomic bomb, releasing into the atmosphere not hundreds of pounds of plutonium but *tons* of the universe's most deadly substance, a pound of which could kill the world's population.

CHAPTER THREE

Accident Hazards

What is the worst accident that can happen at a standard nuclear power plant?

A massive escape of radioactive material, spewed into the environment by an explosion. That can happen in a matter of seconds in a "nuclear runaway" or "power excursion" accident. A reactor building is designed like a fortress-prison or tomb for good reason: its inconceivably poisonous contents must be confined at all costs. Unleashed into the environment, just a small fraction of the two tons of radioactive "fission products" in a nuclear power plant (compared to the two pounds of fission product fallout of the Hiroshima atomic bomb) or just ounces of the up to 1,000 pounds of plutonium built up, would wreak immense catastrophe.

Because most of the radioactive poisons in a nuclear plant have a "hazardous lifetime" into millenia, they would continue to release life-killing radioactivity for that long.

Although a meltdown causing a "China syndrome" (what we've generally been told is the worst accident possible) would indeed produce a disaster unlike any the world has known, a nuclear runaway, also called a power excursion, is even worse. Why? Because it involves an extremely rapid and intense rise in the fission level within the nuclear core—a thousand times beyond normal—simultaneous with sudden overheating, melting of the core, and an instant steam explosion with the power of thousands of pounds of TNT, easily blowing apart the concrete "containment" of a nuclear plant and letting what's inside out.

Here is a photo of a U.S. government test of the explosion potential in a nuclear runaway or power excursion. The test was conducted in 1954 in the Idaho desert on a miniaturized reactor (on a scale of 1:500) with a tiny core, with the reactor operating only a brief time so there was minimal buildup of fission products. The black square in the plume is a piece of equipment weighing a ton.

U.S. Government Photo

Another nuclear runaway occurred in 1961 in a miniature reactor (the SL-1) at the U.S. government's Idaho test grounds. This one was unplanned. Three workers* were killed including one who was found impaled on the ceiling one story above the reactor floor, a reactor control rod through his groin and out his shoulder pinning him to the ceiling. The hands and heads of the victims were so hot with radiation that they had to be severed and buried with radioactive waste. Their bodies were placed in lead-lined caskets and interred in lead-lined vaults.

Here is the government's account of the SL-1 runaway:

*John Byrnes, Richard McKinley and Richard Legg.

SL-1 EXCURSION
Idaho Falls, Idaho, Jan. 3, 1961

A nuclear excursion occurred within the reactor vessel, resulting in extensive damage of the reactor core and room, and in high radiation levels (approximately 500-1,000 rem/hr) within the reactor room.

At the time of the accident, a three-man crew was on the top of the reactor assembling the control rod drive mechanisms and housing. The nuclear excursion, which resulted in an explosion, was caused by manual withdrawal, by one or more of the maintenance crew, of the central control rod blade from the core considerably beyond the limit specified in the maintenance procedures.

Two members of the crew were killed instantly by the force of the explosion, and the third man died within two hours following the incident as a result of an injury to the head. Of the several hundred people engaged in recovery operations, 22 persons received radiation exposures in the range of three to 27 rem gamma radiation total-body exposure. The maximum whole-body beta radiation was 120 rem.

Some gaseous fission products, including radioactive iodine, escaped to the atmosphere outside the building and were carried downwind in a narrow plume. Particulate fission material was largely confined to the reactor building, with slight radioactivity in the immediate vicinity of the building.

The total property loss was $4,350,000. (*See* TID-5360, Suppl. 4, p. 8; 1962 *Nuclear Safety,* Vol. 3, #3, p. 64.) *

*From "Operational Accidents and Radiation Exposure Experience," U.S. Atomic Energy Commission, April, 1965.

How does a nuclear runaway happen?

In a number of ways including accidental withdrawal or ejection of control rods, displacement of the core by mechanical failure or earth movement, a malfunctioning valve.

A nuclear runaway or power excursion is sometimes termed a "reactivity accident" because it creates a sudden, intense jump in the rate of nuclear reaction. When this happens fission erupts in "exponential" growth instead of being controlled. With each act of fission taking place in thirty millionths of a second, it takes but a fraction of a second for the core to become white-hot, melt, and set off a steam explosion ripping apart the containment.

There are two emergency systems in a nuclear power plant: the "SCRAM" system and "emergency core cooling system." SCRAM is supposed to automatically insert the control rods into the core to stop the fission process at the first hint of trouble. But this must be done within one second to stop a nuclear runaway, and SCRAM systems have been found to be inoperable. The "emergency core cooling system" can be used only for a loss-of-coolant accident (in which fission product "decay heat" is the problem). Nothing can cool a nuclear runaway.

Have the government and nuclear industry dealt with this matter? No. They have suppressed information instead.

Nuclear engineer Richard Webb has uncovered a secret report of the U.S.'s National Reactor Testing Station, run by the Phillips Petroleum Company, on the matter of nuclear runaways or the "reactivity accident."

It was 1964 "at a crucial juncture in nuclear power development," notes Dr. Webb. Until that time only small reactors had been built but large nuclear power plants were being designed and proposed to go "on line." The National Reactor Testing Station analyzed the potential for a nuclear runaway in such large nuclear plants; its report concluded that such accidents were possible and would be "catastrophic."

The report recommended:

c. Destructive Tests Involving Essentially Full-Scale Operating Power Reactor Systems.

These "demonstration" tests have three major justifications: to provide information for development of analysis; to climax the complete reactor research program as a test of ability to predict results; and to demonstrate the actual consequences of a reactor accident. The program cannot be considered complete until this type of test is done.

This is an internal Phillips report, and must be handled on that basis. Management approval must be obtained for its distribution outside Phillips Petroleum Company.

PTR-738

A REVIEW OF THE GENERALIZED REACTIVITY ACCIDENT FOR WATER-COOLED AND -MODERATED, UO_2-FUELLED POWER REACTORS

And it declared:

In submitting the following program recommendations, it is considered that time is of the essence, not only because of the large number of present and proposed power reactors of the type considered in this review, but because of the inherent amount of time required to complete a program of this magnitude.

"We should have gone that route," says Dr. Webb, "followed the recommendations and fully explored the potential for nuclear runaways. Instead the federal government suppressed the document, kept it secret, and went ahead and authorized that year the development of these plants."

The U.S. government was trying to run away from the nuclear runaway.

And, in licensing proceedings for commercial nuclear plants, the government has forbidden the discussion of a nuclear runaway, indeed of any major accident type other than certain kinds of loss-of-coolant accidents. The government's justification is that these serious accident types are not, in the government's terminology, "design-basis accidents"—accidents nuclear power plants are designed to handle.

How does a loss-of-coolant accident happen?

Once fission begins in the large mass of fissionable material in a nuclear power plant, heat builds up. Left alone the fuel would get hotter and hotter until, at 5,000 degrees Fahrenheit, it would melt and bore through the steel pressure vessel and the concrete floor of the plant. Two mechanisms *prevent* overheating: the control rods which regulate the level of fission and the water circulating under pressure amid the fuel rods. In the absence of either constraint, the fuel would heat to its melting point. It's as if, while driving, you had to keep your foot on the brake all of the time in order to keep your car from surging ahead and crashing.

But even if the control rods are jammed into the core to stop fission—as occurs in a SCRAM—this alone would not stop the heat buildup. This is because even when fission stops there is what is called "after heat" or "decay heat"—only one to seven per cent of the heat in a reactor during fission but still enough, in the absence of coolant, to melt the fuel in from three to twenty minutes.

The first few seconds after a loss of coolant are vital.

If during the span of fifteen to thirty seconds the emergency core cooling system doesn't function, doesn't send water to the core, a meltdown is unstoppable. Then the "China syndrome" begins. The fuel interacts with the groundwater and radioactivity billows up from the earth around the plant. The white-hot fuel can also interact with water within the plant system and cause steam explosions which, in turn, cause a "breach" of containment and the release of the radioactive poisons inside the plant into the environment. Both "China syndrome" and in-plant containment breaches can occur in the same loss-of-coolant accident.

Emergency core cooling systems have failed many tests.

But even if it does function within fifteen to thirty seconds, the emergency core cooling system can be ineffectual against a meltdown. If the heat has caused the nuclear fuel rods to "crumble"–to fall in a pile on the floor of the reactor vessel–it is impossible for coolant to circulate within the glob to cool it. And a break in the reactor vessel would render any emergency re-flooding useless.

The loss of usual coolant can be caused by a break anywhere in the maze of piping through which a nuclear plant's huge amount of coolant water flows. This water is under high pressure; that is the reason for so much concern about faulty welding on these "pressure pipes." A faulty weld–and they have been common in nuclear plants–can cause the coolant water to burst out as steam (because it is over 500 degrees Fahrenheit and under intense pressure). Such an eruption is called a "blowdown."

In a nuclear runaway, the "full inventory" of radioactive materials in a nuclear plant can be ejected in seconds. In a loss-of-coolant accident less radioactive matter might be thrown into the environment (but because it takes just a small fraction of these poisons to cause huge catastrophe this is little relief) and it would take a little longer.

Other types of nuclear plant accidents are the power-cooling mismatch accident and spontaneous reactor-vessel-rupture accident.

How does a power-cooling mismatch accident happen?

Here a part of the core is producing more heat than the coolant is removing. Ways this can happen include a foreign object blocking a coolant pipe and uneven placement of control rods. Such an accident can cause a series of fuel rod meltings, steam explosions and vessel rupture and set off a nuclear runaway.

How does a spontaneous reactor-vessel-rupture accident happen?

It can occur from cracks developing in the reactor vessel walls because of faulty manufacture or design, or faulty operation, or by failure of the bolts (each supposed to withstand one million pounds of force) which hold down the vessel's lid. In this case, the reactor literally blows its top, the lid flies through the containment and the exposed fuel and fission products vaporize directly into the atmosphere.

A catastrophic accident at a commercial nuclear power plant would have consequences far greater than any single incident in war or peace in the history of the world, an exhaustive, long-secret study made by the U.S. government in the 1960's conceded.

The study was conducted at the government's Brookhaven National Laboratory in New York. Scientists there quickly realized what was involved. Here are sections of their report, the "WASH-740-update":

Dr. Beck asked if the computer programs were ready. Mr. Downes replied that they were running, but the results were frightening. Dr. Winsche noted that unless some mechanism can be found to make their assumptions impossible, the numbers look pretty bad.

The casualties and damage projected by "WASH-740-update" for a catastrophic nuclear plant accident–what the U.S. government lists as a "Class 9" event–are 45,000 deaths, 100,000 injuries and property damage ranging from $17 billion to $280 billion, "an appreciable fraction of the gross national product."

The sequence of calculations that led to our belief that under the worst conceivable conditions, damages could become an appreciable fraction of the gross national product is as follows. Clearly the greatest damage to the population would arise from a release of fission products from a reactor having the greatest fission product inventory. Such an inventory is directly proportional to thermal power and builds up gradually with increase in the lifetime of fuel in the reactor.

Accidents will happen, the government acknowledged.

In any machinery as complex as a reactor facility, it is inevitable that structural failures, instrument malfunctions, operators' errors and other mishaps will occur, despite the most careful design and rigid schedules of maintenance. Such has been the experience with reactor installations. At one Commission installation where seven or so reactors are located, a procedure has been in effect for several years which requires that if any minor mishap or abnormal incident occurs, a brief, formal report must be submitted for the record. Over a period of seven years more than 3,000 such incidents have been recorded. Similarly, as an incidental by-product of a study on the characteristics and qualifications which should be possessed by reactor operators, a list of some 1400 minor incidents was developed through inquiry at 30 different reactor installations.

OFFICIAL USE ONLY

A small opening in the containment would allow the radioactive poisons to get out—fast.

An opening the size of a door, will have an exhaust time due to wind action which is short compared to the fission product deposition time. Under these conditions, most of the fission products would be released to the atmosphere.

Emergency systems can't always be expected to work, the U.S. government admitted.

The Emergency Core Cooling system cannot be made fool proof. It must be turned on and must have an adequate water supply in order to operate effectively. Thus if one of the major coolant pipes fails and the emergency core cooling system also does not perform adequately, then the fuel element temperature would rise, the elements would melt and the fission products would be released from the fuel matrix. An aerosol of fission products could be swept out of the vessel and into the containment shell by convection currents.

"FOR OFFICIAL USE ONLY"

It would all happen quickly, said the government.

FUEL MELTDOWN TIME MODEL

TIME	PHENOMENA	AVE. FUEL TEMP.°F	ENVIRONMENT	IODINE RELEASED
0-10 SEC.	BLOWDOWN	<2000	STEAM-WATER	0
10-1000	INITIAL HEATUP	~2500	STEAM	0
360-1800	CLAD MELTS	2500	STEAM	↑
1800-5400	CORE SLUMPS	2500	STEAM	1.6 %
$1\frac{1}{2}-3\frac{1}{2}$ HRS.	VAPORIZATION OF WATER	<2500	BOILING WATER	↓
$3\frac{1}{2}-4$	BOTTOM HEAD HEATUP	3100	EXPANDING STEAM	10.4 %
>4	BOTTOM HEAD MELT-THROUGH	3500	MOLTEN STAINLESS STEEL	28 %
			TOTAL	40 %

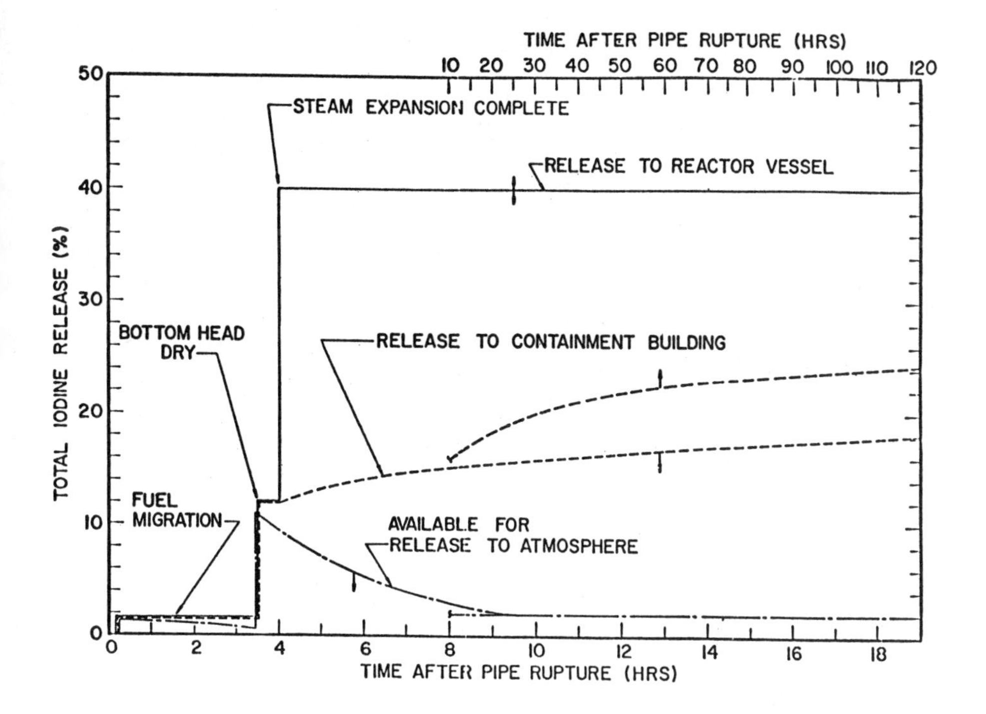

The first two hours would produce the most damaging radioactivity:

Mr. Smith said that the release model kept track of both time and concentration, and that most of the damage is done by that activity released in the first two hours. As a result there were only about 2 or 3 hours available for evasive purposes. Miss Court noted that close-in there was very little that could be done, since doses were higher and received sooner. Mr. Smith added that the population involved is large and would make evasive measures difficult. Dr. Winsche noted that shelter might be helpful, but Mr. Downes said that this probably was not the case, if the air turnover rate in a house was as high as BNL thought it to be.

Radioactivity would be spread far and wide.

> I. Assuming coolant loss in a large 5200 Mwt reactor, a fuel cycle of 1000 days, and sequential failure of all engineered safeguards including the containment structure, radioactive contamination to significant activity levels would be distributed over an area from 10,000 to 100,000 square kilometers. For I^{131}, a contamination level higher than 10 rad to the thyroid would extend beyond 1000 kilometers.

Said the WASH-740-update:

> The result if a city were involved would be catastrophic and there would be deaths out to 150 km.

Even cities at a distance from reactors are not safe in the event of a catastrophic accident because radioactivity blows in the wind.

> conditions would be substantially the same, whether the reactor is in a large city or some distance away, assuming that in the latter case the wind is directed toward the city in the time period considered. The advantage of a "country" location for a large reactor is related to less severe accidents than the extreme ones of case III.

Considering the extent of radioactive contamination, evacuation would not help.

> they would use the latest UN number of 1 or 2 per million per rad to get the number of cancers produced. This will be a large number and may be comparable to the acute results. Mr. Downes noted that evacuation had been considered and was found to make little difference in the results.

With the WASH-740-update complete, the government team which made the study considered what to do with it.

> Dr. Beck indicated that the results of the BML study indicate that the original WASH-740 assumptions cannot be reduced or ignored for liability definition and now we have the problem of how best to present the new information in view of the increase in reactor sizes and the relative increase in potential damage.

The record of another meeting:

> Now that Brookhaven has done the study and gotten results or conclusions, there remained to decide the manner of publishing the BNL report, with discussion but without quantitative results. Dr. Cowan asked how this could be done without anybody (e.g., JCAE) knowing that the results are 50 to 100 times worse. Dr. Beck said that their awareness of the fact that it is worse made the matter of the form of the report very important.

In the end, the report was kept secret.

What is the worst accident that can happen at a breeder reactor?

A breeder reactor can explode like a small atomic bomb. Such a detonation would vaporize and blast out into the atmosphere not only a breeder's full inventory of fission products but the tons upon tons of plutonium with which it is fueled and which it produces.

In a standard nuclear power plant, a nuclear runaway or power excursion would trigger a *steam explosion* blowing apart the concrete containment and letting loose the radioactive poisons inside. In a breeder reactor, it is a *nuclear explosion* that's involved which can be set off by a nuclear runaway, loss-of-coolant or other types of accidents.

To do "breeding"–turning uranium into additional plutonium fuel through "fast" neutron action–the tons of plutonium with which a breeder is fueled must be far more highly concentrated than the uranium-235 fuel in a standard plant. The plutonium in a breeder is at fifteen per cent concentration level instead of the three per cent enrichment level of uranium fuel in a standard plant. Further, the plutonium fuel rods in a breeder must be packed much closer together than the uranium in a standard "light water" reactor.

Thus, there is far more than enough concentrated fissionable material in a breeder reactor to cause an atomic explosion–if it comes any closer together.

The word used for this much-feared event in a breeder is "compaction." If the fuel rods of concentrated plutonium in a breeder reactor buckle or as it is termed, "bow," or melt towards each other, or if there is fuel motion, there can be an atomic bomb-like detonation, blowing any reactor building apart.

Adding further to the volatility of a breeder is the coolant that must be used to maintain fast neutron action: molten sodium, which reacts violently on contact with air or water. A loss or blockage of the liquid sodium in a breeder, even in just a small section of the core, can lead in seconds to melting and compaction of the plutonium fuel and a nuclear explosion.

Even such a nuclear booster as Edward Teller, the "father" of the H-bomb, warns: "For the fast breeder to work in its steady-state breeding condition you probably need something like half a ton of plutonium. In order that it should work economically in a sufficiently big power-producing unit, it probably needs quite a bit more than one ton of plutonium. I do not like the hazard involved . . . . If [nuclear reactors] malfunction in a massive manner, which can happen in principle, they can release enough fission products to kill a tremendous number of people . . . . If you put together two tons of plutonium in a breeder, one tenth of one per cent of this material could become critical . . . . Although I believe it is possible to analyze the immediate consequences of an accident, I do not believe it is possible to analyze and foresee the secondary consequences. In an accident involving a plutonium reactor, a couple of tons of plutonium can melt. I don't think anybody can foresee whether one or two or five per cent of this plutonium will find itself and how it will get mixed with some other material. A small fraction of the original charge can become a great hazard."*

A single large breeder reactor would "contain the equivalent of about 10,000 to 20,000 atomic bombs of total plutonium radioactivity," notes Dr. Richard Webb. After a breeder accident, he calculates, the "abandonment area" (based on U.S. government standards of allowable levels of plutonium in the environment) for the kind of large breeders planned for the U.S. and Europe, comes to 360,000

*From paper by Teller presented to the New York section of the American Nuclear Society, 1967.

square miles. That is forty per cent of the land east of the Mississippi River.

"Fast breeder reactor cores are prone to mishap," says Dr. Webb in an analysis conducted on the SNR-breeder reactor under construction near Kalkar, West .Germany. Experience bears this out: of the three small fast breeders built in the U.S., two have had near-catastrophic fuel melting accidents.

- America's first "experimental breeder reactor," the EBR-1, suffered a partial melting of its fuel in 1955 and never operated again. Subsequent studies showed the reactor was but a half-second away from an A-bomb-like blast. The reactor, with a miniature core and sited in the Idaho desert, was being brought up in power when the fuel rods began to bow inwards towards each other. Here is a U.S. government photo of the devastated core:

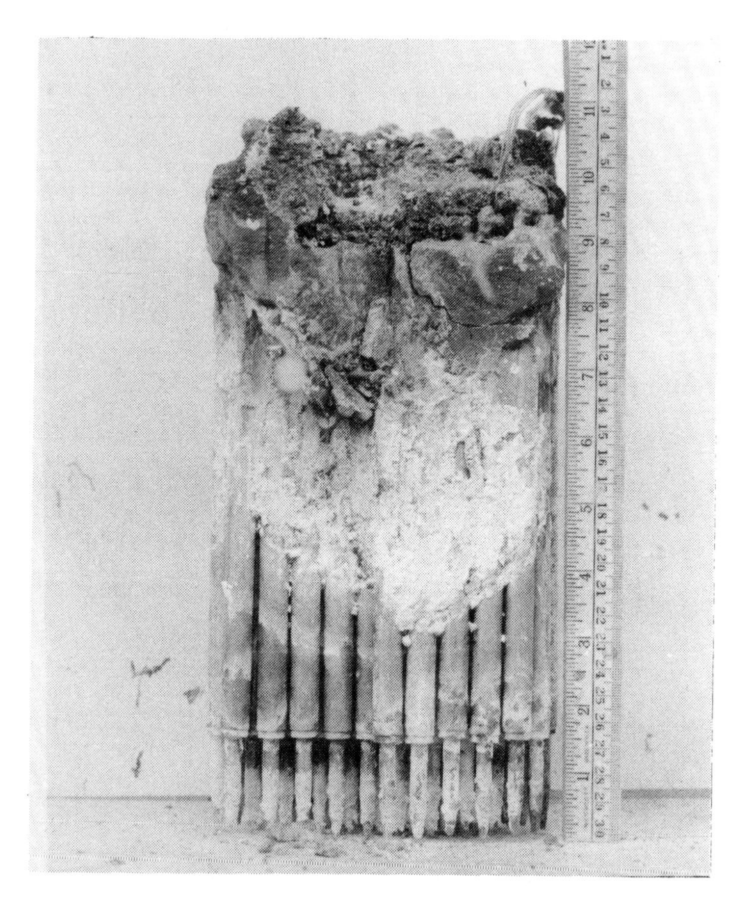

• The Enrico Fermi Power Plant at Lagoona Beach, Michigan, twenty miles from Detroit, was America's first and so far only "demonstrator" commercial breeder reactor. In 1966 it underwent a partial fuel meltdown caused by some loose pieces of metal blocking the coolant flow. For weeks engineers fought a touch-and-go battle to get the reactor under control. "Let's face it, we almost lost Detroit," declared one Fermi engineer later, quoted in the book *We Almost Lost Detroit.* A suppressed U.S. government report earlier had warned: "There is insufficient information available at this time to give assurance that the reactor can be operated at this site without public hazard."* And a minority on the U.S. Supreme Court took the view, on a legal challenge to block the plant from going into operation led by the late American union leader Walter Reuther, that the then Atomic Energy Commission was involved in "a lighthearted approach to the most awesome, the most deadly, the most dangerous process that man has ever conceived." Plans to try to re-start the Fermi breeder were abandoned.

In Russia, there was a sodium-water eruption of a breeder on the Caspian Sea in 1973, violent enough to appear on a U.S. reconnaissance satellite.

Still, the push is on to build breeders. Even if they don't generate electricity well, or at all, they are seen as vital to produce plutonium for use as fuel in the standard nuclear plants as fissionable uranium runs out.

The "Review Of National Breeder Reactor Program" contains these passages:

> Our best calculations are that we have enough low-cost uranium to fuel all plants built until sometime in the 1990's, for their entire lifetime. Unless we find vast deposits of high-grade uranium between now and then (preferably within the United States), we will be unable to continue building the type of nuclear power plants we are relying on today.

> We will have only a few breeders on the line by the end of this century. Their primary purpose, however, will be in guaranteeing a fuel supply for the hundreds of ordinary nuclear power plants that will be supplying our energy at that time. Assuring that fuel supply is the most important role for the breeders during this century; supplying electricity is secondary.

*From July 6, 1956 memo from Roger S. McCullough, chairman of the AEC's Committee on Reactor Safeguards, a document kept secret by AEC Chairman Lewis Strauss.

94TH CONGRESS } JOINT COMMITTEE PRINT
2d Session }

REVIEW OF NATIONAL BREEDER
REACTOR PROGRAM

REPORT

BY THE

AD HOC SUBCOMMITTEE TO REVIEW THE LIQUID
METAL FAST BREEDER REACTOR PROGRAM

OF THE

JOINT COMMITTEE ON ATOMIC ENERGY
CONGRESS OF THE UNITED STATES

JANUARY 1976

Printed for the use of the Joint Committee on Atomic Energy

U.S. GOVERNMENT PRINTING OFFICE
58-550 WASHINGTON : 1976

A breeder reactor proponent, Alvin Weinberg, long-time director of Oak Ridge National Laboratory, who continues to have a major say in U.S. nuclear policy as current director of the heavily government-funded Institute for Energy Analysis at Oak Ridge, speaks of 2,000 breeder reactors in the U.S., 10,000 in the world in the next century.

This is despite his projection of a breeder "meltdown every two years," and despite his observation that such a meltdown "anywhere is a . . . meltdown everywhere." Still, he regards plutonium breeders as the preferable form of nuclear power and, because of uranium shortages, thinks breeders should have been built in the first place instead of uranium-fired nuclear plants. To cope with the dangers he proposes that the breeders be built in clusters "in sparsely settled areas" with extreme security. "In a sense we will buy order at the cost of freedom," he says.*

*Address of Dr. Weinberg at Brookhaven National Laboratory, May 17, 1977.

Can an atomic bomb-like explosion occur at a standard nuclear power plant?

This is in dispute. Such a nuclear explosion had been considered impossible because of the low (three per cent) concentration of uranium nuclear fuel. But recent analysis by Dr. Webb points to the potential of the 1,000 pounds of plutonium built up in a standard plant concentrating in a critical mass during a meltdown and exploding.

What have been some of the serious nuclear power plant accidents so far–besides the Fermi, SL-1 and EBR-1 accidents?

THREE MILE ISLAND. Here, in 1979, the failure of a "main feedwater" system set off a series of failures–of faulty gauges and valves, of operator error and zirconium flare-up–all confirming that "Murphy's Law," the credo of engineers, that if something can go wrong, it will, operates on a broad scale amidst the complexity of a nuclear power plant. At this Pennsylvania plant there was a loss-of-coolant accident, a hydrogen explosion, and an escape of radioactivity into the atmosphere. The plant narrowly escaped a meltdown that would have led to the magnitude of disaster that the WASH-740-update declared "might be equal to that of the State of Pennsylvania." Dr. Webb, in his extensive analysis of the accident, points out that given the multiple failures only multiple "strokes of luck" prevented such catastrophe: luck that the hydrogen bubble, formed as the zirconium fuel cladding erupted, finally dissipated; luck that the reactor had been in operation for only three months so the decay heat was minimal; luck that a critical mass of plutonium did not form; luck that "nothing else failed considering the many failures that occurred."

Says Webb: "The accident should be clear warning that multiple-failure accidents are likely to occur," although U.S. government nuclear officials have only been concerned about "single failure" mishaps.

As the President's Commission on the Accident at Three Mile Island reported, there will have to be "fundamental changes" made in how nuclear power is supervised, but even then "we do not claim that our proposed recommendations are sufficient to assure the safety of nuclear power."

Report Of
The President's Commission On

THE ACCIDENT AT THREE MILE ISLAND

The Need For Change:
The Legacy Of TMI

OVERVIEW

OVERALL CONCLUSION

In announcing the formation of the Commission, the President of the United States said that the Commission "will make recommendations to enable us to prevent any future nuclear accidents." After a 6-month investigation of all factors surrounding the accident and contributing to it, the Commission has concluded that:

> To prevent nuclear accidents as serious as Three Mile Island, fundamental changes will be necessary in the organization, procedures, and practices -- and above all -- in the attitudes of the Nuclear Regulatory Commission and, to the extent that the institutions we investigated are typical, of the nuclear industry.

This conclusion speaks of *necessary* fundamental changes. We do not claim that our proposed recommendations are sufficient to assure the safety of nuclear power.

The "Special Inquiry Group" report commissioned by the NRC declared about the TMI accident:

engineering calculations performed during our investigation indicate that on the morning of March 28, before anyone appreciated the seriousness of the situation, Three Mile Island came close to being the accident we had been told by many in the industry could not happen: a core meltdown. A shift foreman reporting for normal duty about 2 hours after the accident began undertook to survey some instruments and blocked off the stuck-open pressurizer valve that was leaking reactor coolant into the reactor containment building. If that block valve had remained open, our projections show that within 30 to 60 minutes a substantial amount of the reactor fuel would have begun to melt down—requiring at least the precautionary evacuation of thousands of people living near the plant, and potentially serious public health and safety consequences for the immediate area.

We were not asked, and it is not our place to tell the public, "how safe is safe enough." Indeed, as we make clear in this report, we believe this is a decision that in the final analysis should not be the exclusive province of the NRC: it is an executive decision that should be made as a part of our national energy strategy by the Executive and by Congress. The NRC cannot continue to face, *sub silentio,* in every policy and licensing determination the question of the future of nuclear power in this country. It is, lest we forget, an inherently dangerous activity that Congress has authorized the NRC to license.

The generation of nuclear power can never be risk-free. It will inevitably present certain risks to public health and safety no matter how "safe" plants are made.

We have found an industry in which the expertise and responsibility for safety is fragmented among many parties—the utility company that operates the plant, the plant designer, the manufacturer of the reactor system, the contractor, and the suppliers of critical components, in addition to the NRC. Coordination among these parties and between them and the NRC, as well as within the NRC, is inadequate. As a result, there are many institutional disincentives to safety, and safety issues that are identified at some point in the system often fall through the cracks. Prior to Three Mile Island, the industry as a whole had made only feeble attempts to mount any industrywide affirmative safety program, and many utilities apparently regarded bare compliance with NRC minimum regulations as more than adequate for safety.

BROWNS FERRY. In 1975, the system of electrical cabling converging in the common control room of the largest nuclear facility in the world, twin Tennessee Valley Authority nuclear plants in Alabama, burst into flames when workmen used a candle to test for air leaks. The blaze quickly spread and burned uncontrolled through the facility for seven and a half hours, destroying or incapacitating plant safety systems including control of reactor coolant systems and the emergency core cooling system in one unit. The reactor water level plummeted, almost uncovering the core in that unit; a meltdown was projected as an hour away. Operators fought to gain control of the facility and as one TVA engineer said later, did so by "sheer luck." It was shut down for repairs, which cost $150 million, for eighteen months.

LUCENS REACTOR. This thirty megawatt reactor built inside a mountain in Switzerland underwent a loss-of-coolant accident in 1969. The cavern which held the reactor was sealed.

WINDSCALE. Large quantities of radioactive iodine were released over northern England when graphite core material in this nuclear facility, on the Irish sea, caught fire in 1957. Dairy farming was banned for sixty days over a 200 square mile area, after milk samples showed six times the permitted limit of radioactive iodine. Cattle had to be slaughtered in the 200 square mile area and their thyroid glands were removed. A vast radioactive cloud moved over London, 300 miles away.

CHALK RIVER. A defective fuel rod improperly removed caused this experimental reactor in Canada to come close to a meltdown in 1952. Large quantities of radioactive material were released.

The list goes on and on. To mention just a few more:

OYSTER CREEK reactor in New Jersey had a series of multiple failures one month after the Three Mile Island accident. The reactor coolant level dropped to one foot above the core before malfunctioning valves could be re-opened.

DRESDEN reactor in Illinois went out of control in 1970. For two hours radioactive iodine was released.

VERMONT YANKEE went "critical" during a maintenance operation in 1973 and came close to a nuclear runaway.

MILLSTONE, a nuclear facility in Connecticut, has undergone a long series of radioactive leaks and explosions.

DAVIS-BESSE in Ohio in 1978 and CRYSTAL RIVER in Florida in 1980 had accidents that were similar to TMI.

"The question is frequently asked," says Dr. Webb, who served under Admiral Hyman Rickover on America's first large-scale nuclear power plant, the government-owned Shippingport facility near Pittsburgh, "Can nuclear power plants be redesigned and made safe?"

"The answer is no," says Dr. Webb who went on to specialize in nuclear plant accident potentials and consequences, receiving a Ph.D. and becoming an authority in the field.* "Firstly, adding additional safety equipment and operator instructions makes nuclear plants even more complex than they already are. More complexity could increase, not reduce, the probability of accidents. . . . Secondly, a major redesign of reactors would not yield safe reactors. Nuclear reactors by their very nature will always have the potential for power excursion accidents, and loss-of-coolant accidents. Reactors cannot avoid making the radioactive by-products, and there will always be decay heat source after the fission heat is stopped.

"Experience shows us that reactor accidents are occurring regularly and each time they are getting worse and worse. . . . We are forced to conclude on the basis of experience and the potential for catastrophic accidents and the virtually infinite number of ways such accidents can occur," says Dr. Webb, "that an accident which potentially can result in a public catastrophe is likely to occur."

He predicts it happening in the 1980's.

As Russell Peterson, a member of the President's Commission on the Accident at Three Mile Island, declared in his "supplemental" view to the group's report:

> As a final comment, I wish to emphasize my conviction, strongly reinforced by this investigation, that the complexity of a nuclear plant--coupled with the normal shortcomings of human beings so well illustrated in the TMI accident--will lead to a much more serious accident somewhere, sometime. The unprecedented worldwide fear and concern caused by the TMI-2 "near-miss" foretell the probable reaction to an accident where a major release of radioactivity occurs over a wide area. It appears essential to provide humanity with alternate choices of energy supply. Accordingly, I recommend the development by our federal government before we become more fully committed to the vulnerable nuclear energy path, of a strategy which does not require nuclear fission energy.
>
> Russell W. Peterson
>
> October 25, 1979

*See Webb's landmark book, *The Accident Hazards of Nuclear Power Plants,* The University of Massachusetts Press, Amherst, Mass., 1976.

The full Commission report on Three Mile Island declared:

While throughout this entire document we emphasize that fundamental changes are necessary to prevent accidents as serious as TMI, we must not assume that an accident of this or greater seriousness cannot happen again, even if the changes we recommend are made. Therefore, in addition to doing everything to prevent such accidents, we must be fully prepared to minimize the potential impact of such an accident on public health and safety, should one occur in the future.

To deal with the expected nuclear catastrophes ahead preparations have been made—quietly, without your knowing about them yet—which essentially consist of having you run or hide.

PROTECTIVE ACTION EVALUATION

PART I

THE EFFECTIVENESS OF SHELTERING AS A
PROTECTIVE ACTION AGAINST NUCLEAR
ACCIDENTS INVOLVING GASEOUS RELEASES

APRIL 1978

George H. Anno
Michael A. Dore

Prepared for

U.S. Environmental Protection Agency
Office of Radiation Programs
Washington, D.C. 20460

IV. CONCLUSIONS AND RECOMMENDATIONS

Shelter protection provided by a large variety of public structures can provide a significant reduction in WB and thyroid dose from exposure to radioactive gaseous fission products that might be released during a nuclear power plant accident. Protective sheltering is attractive if shelter-access timing is ideal, but its effectiveness diminishes almost linearly with access delay time after cloud arrival.

Sheltering protection against inhalation exposures that result in thyroid dose depends on the number of air changes taking place over the period of exposure to airborne radioactive cloud material. Sheltering protection for WB exposures depends on the attenuation of gamma radiation originating from the airborne cloud source, the number of air changes during cloud exposure, and (to a lesser extent) the attenuation of gamma radiation originating from the ground fallout about the shelter structure. Accordingly, optimum ventilation control (low air-change rates during cloud passage) is more effective for reducing thyroid dose than WB dose. Albeit, ventilation control is relatively more effective for reducing WB dose in LS than in SS.*

Large structures such as office buildings, multistory apartment complexes, department stores, etc., generally would provide significantly more sheltering for WB exposures than smaller structures such as single-family dwellings--a factor of about 4.5 more during low air-change rate conditions and 3 more for nominal air change rates. That is, WB doses would be reduced by a factor of 2.5 to 3 for SS sheltering; whereas for LS sheltering, WB doses would be reduced by a factor of about 12 during low air-change rate conditions. For representative air change rate conditions, WB dose would be reduced by about 2.3 for SS and from 6 to 9 for LS. WB dose can be further reduced in a shelter structure through use of expedient filtration; e.g., by stuffing cracks and openings with cloth or paper materials, which would reduce radioactive material ingress (discussed above, p. 10 ff.) and/or the natural ventilation rate. Similarly, another means of respiratory protection is to cover the nose and mouth area with such common items as towels, handkerchiefs, or toilet paper: e.g., a crumpled handkerchief (or one with eight or more folded layers), a towel of three or more folded layers, or toilet paper of three or more folded layers can reduce inhaled radioactive material (particulate

*(WB–whole body dose radiation exposure; LS–large structures; SS–small structures.)

iodine in this study) by a factor of about 10 [35]. The reduction of WB dose in a SS, however, is not appreciable--about 2.5 percent for low ventilation rates and about 15 percent for representative ventilation rates. The reduction in WB dose in a LS would be more appreciable--about 13 percent for low ventilation rates and about 70 percent for representative ventilation rates.

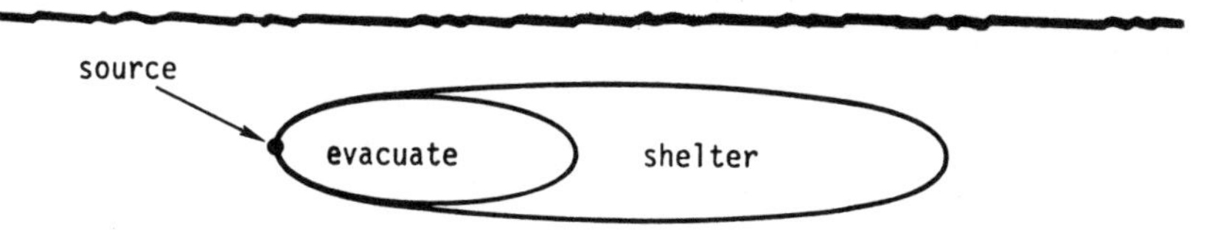

Note, however, that for the above consideration the high-dose area would normally be closest to the origin of release for ground-release assumptions. Accordingly, time constraints could limit the number of individuals who could be evacuated effectively.

Finally, there will be cases where LS sheltering gives adequate protection and is cheaper and quicker than evacuation, but where it is desirable to evacuate individuals for whom only SS shelter is available. This sort of shelter-availability split may be appropriate because timely evacuation may be more difficult in areas where LS are more prevalent than SS.

In summary, for emergency planning purposes, evacuation potentially provides the greatest margin of protection and should be the primary means of protective emergency action in the event of a gaseous fission-product release. On the other hand, sheltering may be the recommended protective action for two reasons: 1) because it is probably less expensive and less disruptive of normal activity than evacuation, sheltering may be appropriate under conditions of marginal danger; 2) it may be more expedient than evacuation. Since the majority of people are already inside some sort of shelter most of the time, mobilization time would be shorter, and the information that must be transmitted to them may be simpler than a set of evacuation instructions. Sheltering, therefore, may be appropriate if the time before cloud arrival is short, even though subsequent evacuation may be desirable.

Additional work is needed to develop complete guideline information for evacuation and sheltering recommendation procedures. For sheltering, both experimental and analytical areas are identified that would lead to the more accurate assessment of protection provided by available

If evacuation or sheltering cannot be effected until after the cloud has passed, individuals in the open would have contracted most of their dose (assuming they do not linger in the open ground-contaminated area more than a few hours after cloud passage) and any additional actions would be dictated by the urgency of avoiding additional exposure from ground fallout or ingestion; e.g., control measures that may very well include subsequent evacuation. However, the actual delay time, T_D, may be less for sheltering than for evacuation, in which case sheltering could provide some degree of protection depending on the access time required during exposure to an airborne cloud and fallout source. As defined in this study, the delay time for evacuation, T_D, is consistent with that defined by EPA, which includes the mobilization time estimated to be from 0.2 to 2 hr for evacuation [8]. For sheltering, the mobilization time and, hence, the corresponding delay time could be significantly less, resulting in a smaller time delay for shelter access. Summarizing, emergency planning guidance when evacuating can either be 100-percent effective (accomplished prior to cloud arrival) or virtually ineffective (could not be effected until after cloud passage):

1. If the projected dose exceeds the PAG by more than a few-fold, and
   a. timely evacuation is feasible, then evacuation is recommended; or
   b. timely evacuation is not feasible (i.e., the time available before cloud arrival is short compared with the required mobilization, warning, and transit time for evacuation), then sheltering is recommended.
2. If the projected dose does not exceed the PAG by more than a few-fold, then sheltering will probably be adequate and economically preferable.

Insofar as sheltering doses limit the degree of protection that can be offered by structures normally available to the public, but still can offer significant protection advantages, a condition--dictated by the time and logistic considerations--exists for which it is necessary to examine both evacuation and sheltering tradeoff options. For this condition, no all-encompassing statements of rule can be made if the optimum

PLANNING BASIS FOR THE DEVELOPMENT OF STATE AND LOCAL GOVERNMENT RADIOLOGICAL EMERGENCY RESPONSE PLANS IN SUPPORT OF LIGHT WATER NUCLEAR POWER PLANTS

A Report Prepared by a
U. S. Nuclear Regulatory Commission and
U. S. Environmental Protection Agency
Task Force on Emergency Planning

H. E. Collins* B. K. Grimes**
Co-Chairmen of Task Force
F. Galpin***
Senior EPA Representative

Manuscript Completed: November 1978
Date Published: December 1978

*Office of State Programs
**Office of Nuclear Reactor Regulation
U. S. Nuclear Regulatory Commission
Washington, D.C. 20555
***Office of Radiation Programs
U. S. Environmental Protection Agency
Washington, D. C. 20460

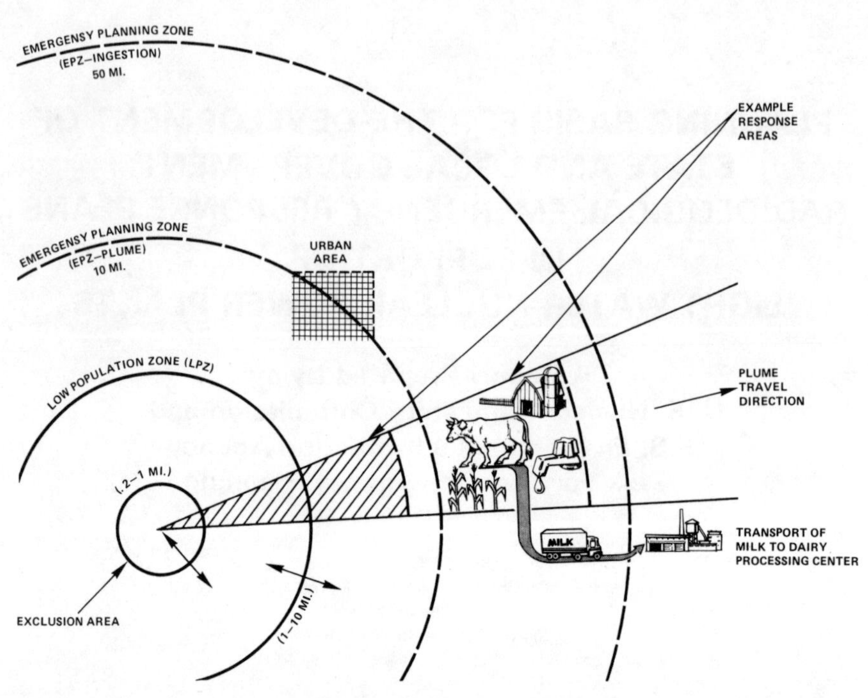

Figure 1 Concept of Emergency Planning Zones

Table 1. Guidance on Size of the Emergency Planning Zone

Accident Phase	Critical Organ and Exposure Pathway	EPZ Radius
Plume Exposure Pathway	Whole body (external)	about 10 mile radius*
	Thyroid (inhalation)	
	Other organs (inhalation)	
Ingestion Pathway**	Thyroid, whole body, bone marrow (ingestion)	about 50 mile radius***

NUREG-75/111
(Reprint of WASH-1293
Revision No. 1 12-01-74)

Guide and Checklist for Development and Evaluation of State and Local Government Radiological Emergency Response Plans in Support of Fixed Nuclear Facilities

U. S. NUCLEAR REGULATORY COMMISSION
OFFICE OF INTERNATIONAL AND STATE PROGRAMS

L. MEDICAL AND PUBLIC HEALTH.

PLANNING OBJECTIVE

To determine the availability within the State and local communities of public and private medical facilities that could accommodate and care for persons involved in a radiological emergency who may require medical care, and to establish the role of each medical facility in the medical and public health support plan. (See footnote 1.)

GUIDANCE	PLAN REFERENCE
1. Public, private, and military medical and first-aid support facilities within the State and local communities (including those of the nuclear facilities) capable of providing both emergency and definitive care of offsite victims of a radiological accident should be identified.	
2. A medical response plan for dealing with nuclear facility incidents with offsite consequences should be developed.	
3. Maps showing the physical location of all public, private and military hospitals and other emergency medical services facilities within the State considered capable of providing medical support for any offsite victims of a radiological incident should be included in the plan. These emergency medical services facilities should be able to radiologically monitor	

Footnote 1.

The availability of an integrated emergency medical services system and a public health emergency plan serving the area in which the facility is located and, as a minimum, equivalent to the Public Health *Service Guide for Developing Health Disaster Plans, 1974* and to the requirements of an emergency medical services system as outlined in the *Emergency Medical Services System Act of 1973 (P.L. 93-154)*, should be a part of and consistent with overall State or local disaster control plans and should be compatible with the specific overall emergency response plan for the facility.

contaminated personnel, and have facilities and trained personnel able to care for victims of radiological accidents. Persons may also need to be evacuated to a hospital due to an existing physical condition not related to the incident.

4. Arrangements should be developed for transporting offsite victims of radiological accidents to medical support facilities.

5. A system for State public health medical recording and followup of radiologically exposed individuals should be established in collaboration with the local or State medical association. The record should include such items as location at time of emergency, radiation dose, contamination status, treatment status, and release status. (See footnote 2.)

6. Training programs should be developed for medical support personnel who may be called upon to care for offsite victims of a radiological accident.

7. A limited list of qualified medical consultants who can, if required, assist State/local government medical authorities in the event of a nuclear facility emergency, should be developed.

8. Medical facilities and ambulances should be equipped with emergency communications capability for intrasystem communications as well as for communications with the State and local government emergency operating centers.

Footnote 2.

The circumstances under which medical attention would be required or useful for any offsite victims of a radiological accident should be determined by guidance provided by the State or local government public health officer, in consultation with Federal health authorities, private physicians and hospitals.

M. RECOVERY AND REENTRY* PLANNING AND POSTACCIDENT OPERATIONS.

PLANNING OBJECTIVE

To determine that general recovery and reentry* plans for areas surrounding the nuclear facility will complement similar plans established for nuclear facilities. (*Reentry as used here refers to reentry to offsite areas that have been evacuated.)

GUIDANCE	PLAN REFERENCE
1. A State technical group should be established with responsibilities to develop, direct, and evaluate offsite area recovery and reentry operations.	
2. Provision should be made and criteria developed for establishing controlled areas or zones surrounding a nuclear facility accident site including access/egress control provisions and perimeter radiological surveillance of persons entering or leaving areas or zones. Criteria should be developed for decontrolling areas or zones. Criteria should be developed for reentry into homes and evacuated areas.	
3. Provision should be for general postaccident operations, e.g., traffic control plans, emergency personnel relief plans, provision of food and shelter, and transportation for emergency personnel.	
4. The availability of commercial businesses and institutions that can provide technical assistance to State and local governments in recovery or reentry operations should be ascertained.	
5. Provision should be made for the positive control or diversion of food and water supplies that may be radiologically contaminated.	

What about the accident hazards of transporting radioactive materials?

Hundreds of thousands of shipments of radioactive materials, from ounces to tons, now move yearly in trucks over our roads and in airplanes in our skies. The AEC, in a 1967 study, projected that for every 10,000 shipments of radioactive materials in all kinds of carriers there would be an accident.* And accidents have been mounting up. Between 1971 and 1975, the Department of Transportation counted 144 accidents involving nuclear shipments. The worst was a spill of 15,000 pounds of powdered uranium oxide from a truck that overturned in Colorado in October 1977. On airplanes, routine shipments of radioactive material in the hold regularly irradiate unsuspecting passengers.

The Nuclear Regulatory Commission 1980 "interim rules" advise transporters of nuclear materials that "areas of high population density" are "to be avoided."**

At the same time, as communities throughout America pass laws to block the movement of radioactive materials within their boundaries, the federal government is moving to override them. In March 1980, the Department of Transportation announced that it plans to adopt regulations overriding the laws and permitting the shipment of radioactive materials anywhere in the U.S. It argues that federal power over interstate transportation is supreme.

This DOT "rule-making" grew out of a challenge by Brookhaven National Laboratory which has been blocked by a local law from shipping nuclear material through New York City.

When the local law was being considered, Leonard Solon, director of New York City Health Department's Bureau for Radiation Control, spoke of "the streets of New York City having routinely been used for the transportation of radioactive materials." This included movement over midtown Manhattan streets "and bridges to John F. Kennedy Airport for shipment by air into and out of the United States of large quantities of mixed isotopes of plutonium . . . . Dispersion of even a small fraction of the contents of one of these shipments as the result of an air crash, concomitant fire, and high winds within the City of New York . . . could have cataclysmic results bringing death or serious injury to tens of thousands of New Yorkers.

*"The Accident Experiences of the USAEC in the Shipment of Radioactive Material," by D.E. Patterson.

**NUREG-0561, U.S. Nuclear Regulatory Commission, Washington, D.C., 1980.

... There is a certain massive intellectual inertia possessed by large influential scientific organizations staffed by the most competent and well meaning of scientists and engineers, who with ingenuity and scholarship, can rationalize and justify almost anything. This includes the preposterous technical obscenity of plutonium air shipments into New York City."

What about government regulation?

It has been a sham from the outset.

As Robert D. Pollard, who was with the AEC and then the NRC for six years, testified before the House Subcommittee on Energy and Power in July 1978:

> From this personal experience as a reactor engineer and project manager, I believe that the core of this country's nuclear regulatory problems is the fact that the entire process is largely a charade designed to create the appearance of legitimacy. This has come about because the regulatory process has been controlled by single-minded individuals who, in their zeal to promote the growth of the nuclear industry, have deliberately and systematically abdicated their responsibility to regulate nuclear power in the manner necessary to protect public health and safety.

Most regulatory standards have been drafted by the nuclear industry itself. Radiation monitoring is left to the utilities. Only a small fraction of "safety-related" nuclear plant activities are checked by government inspectors who spend most of their time examining samples of utility records.

In *Looking But Not Seeing, The Federal Nuclear Power Plant Inspection Program,** Lawrence S. Tye explains that "the official conception of NRC inspectors as thoughtful guardians of public interests in nuclear affairs quickly dissolves upon closer scrutiny. Federal inspectors devote little time to inspecting reactor facilities directly, instead focusing their efforts on auditing records prepared by plant management, records whose accuracy is essentially taken for granted."

*Union of Concerned Scientists, Cambridge, Mass., December 1978.

The AEC "wanted to encourage a rapid construction program and encouraged a loose regulatory framework" to do that, and the NRC inspection staff "is today comprised almost entirely of former AEC employees. Evidence of the persistent cooperation/collusion between the nuclear industry and government regulators abounds. NRC inspectors' reliance on industry records in performing their review function greatly limits their independence. Too many regulatory criteria are voluntary and therefore unenforceable. The government accords little support to plant employees who raise charges of faulty work practices. Reactor inspectors and Headquarters personnel typically understate the safety significance of inspections and findings, opting for mild enforcement sanctions. Many NRC regulatory personnel have in the past and will in the future work for the nuclear industries they oversee. Little progress has been made since the days of the AEC in divesting NRC of its promotional biases. Further, most regulatory standards were drafted by industry associations and often represent the lowest common denominator of accepted industry practices."

In 1979 the Union of Concerned Scientists obtained under the Freedom of Information Act a collection of documents which the NRC itself called "The Nugget File" involving incredible activities at nuclear plants and incredible responses to them by the AEC and NRC.

One March 1968 document describes the use of "a regulation basketball," wrapped in tape to increase its diameter by two inches, inserted into the intake of a reactor cooling pipe and inflated, for use as a seal. But the basketball was forced out of the pipe by the water pressure of 500 pounds, leaving 1,400 gallons of radioactive water to spill on the floor of the unidentified reactor. The government declared: "Where risks of fuel melting and personnel safety are involved, consultation with knowledgeable people should be made prior to questionable operation."

Here is the document:

EXPERIENCES

REACTOR SAFETY

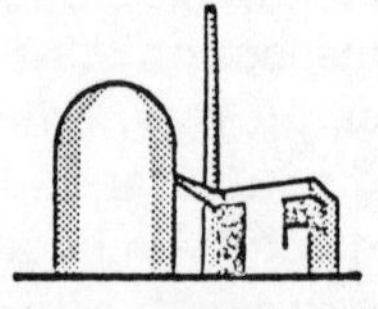

UNITED STATES ATOMIC ENERGY COMMISSION

ROE: 68-1 March 5, 1968

Loss of Pool and Canal Water

1.0 Summary

The shielding water covering spent fuel, capsules and other radioactive material at a test reactor was lowered about six feet when a temporary water seal, installed in the pool pump suction line to permit pump relocation, was expelled from the line. The radiation dose rate at the top of the water increased by a factor of about 130 to 2 Rem/hr. Fourteen thousand gallons of water spilled onto the basement floor and flooded the floor to a maximum depth of about 8 inches.

A regulation basketball, wrapped in rubber tape to increase the diameter by 2 inches, was inserted into the intake of the suction line and inflated. The basketball had been fitted with a wire harness and the harness was fastened to a nearby pipe in the pool. The reactor pool water level was then raised five feet, to the level normally maintained during shutdown, to permit capsule and irradiated component handling in the reactor tank. The ten-inch suction line was parted and work on pump relocation began. Only the basketball plug prevented leakage out of the open suction line. (See attached figure)

5.0 Conclusions

Where risks of fuel melting and personnel safety are involved, consultation with knowledgeable people should be made prior to questionable operation.

U. S. Atomic Energy Commission
Division of Operational Safety

An April 1969 document spoke of radioactivity found in a plant drinking fountain at an unidentified nuclear plant. An investigation led to the discovery that a hose from a well water tap was connected to a 3,000 gallon radioactive waste tank. The government concluded: "The coupling of a contaminated system with a potable water system is considered poor practice in general."

Here is that document:

EXPERIENCES

REACTOR SAFETY

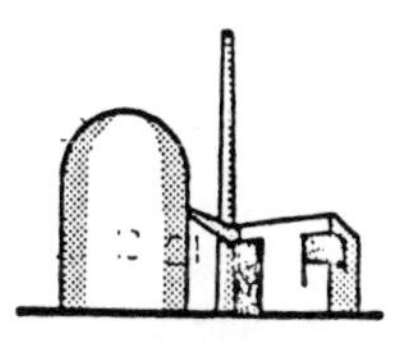

UNITED STATES ATOMIC ENERGY COMMISSION

ROE: 69-10

April 21, 1969

Contamination of Well Water System

1.0 Summary

During a routine check at a power reactor facility, above-normal radioactivity levels were discovered in the well water distribution system for the reactor building.

The contamination is postulated to have resulted from cross connection of a well water tap to a 3000 gallon radioactive waste tank by means of a hose coupling.

2.0 Circumstances

Samples of water taken from taps at a laboratory sink showed radioactivity levels above normal background. Further checking confirmed the presence of radioactivity in one of the plant drinking fountains. Both the sink and the drinking fountain are supplied by the plant well water system.

Investigation included measurement of the radioactivities at intermediate points of the well water system, and led to the discovery of a hose connected from a well water tap to a liquid level transmitter line for a 3000 gallon radioactive waste tank. This discovery, in turn, led to the postulation that radioactive waste was the source of contamination of the well water system.

3.0 Conclusion

The coupling of a contaminated system with a potable water system is considered poor practice in general and should be conducted only when absolutely necessary and under controlled conditions.

Division of Reactor Licensing
U. S. Atomic Energy Commission

Members of the NRC, like those of the AEC before it, huckster nuclear power. Joseph Hendrie, the NRC chairman, in the hours after the Three Mile Island accident declared, "We are operating almost totally in the blind. His information is ambiguous, mine is non-existent and I don't know, it's like a couple of blind men staggering around making decisions."*

Yet a year later he was saying that the U.S. would have 150 to 160 nuclear plants in operation in the 1980's and "this is very useful . . . but it is not large enough to give us the flexibility we need. To obtain that flexibility we will need a new group of nuclear units, of at least the size of the present group and coming into operation from the mid-1990's onward."** (While AEC's deputy director for technical review, Hendrie, a nuclear physicist, wrote a memo—obtained by the Union of Concerned Scientists under the Freedom of Information Act in 1978—which, in replying to recommendations about faults in containment systems declared: "Steve's idea to ban pressure suppression containment systems is an attractive one in some ways . . . . However, the acceptance of suppression containment concepts . . . is firmly imbedded in the conventional wisdom. Reversal of this hallowed policy could well be the end of the nuclear power period.")

And, meanwhile "deaths per gigawatt" are considered.

Possible measures are

Dollar cost per gigawatt-year of electricity generated.

Area and duration of ground temporarily rendered uninhabitable, in hectare-years per gigawatt-year.

Man-rems population exposure per gigawatt-year

Deaths per gigawatt-year

Man-years reduced life expectancy per gigawatt-year†

Plans are being made for the stockpiling of potassium iodide pills to try to offset damage to the thyroid gland by the ingestion of radioactive iodine during a nuclear plant catastrophe. Here is the federal proposal:

*Transcript of NRC deliberations, March 30, 1979.
**Hendrie speech before the Chattanooga Engineering Societies, February 21, 1979.
†From correspondence on U.S. Nuclear Regulatory Commission's WASH-1400 report.

[4110-03-M]

DEPARTMENT OF HEALTH, EDUCATION, AND WELFARE

Food and Drug Administration

[Docket No. 78D-0343]

POTASSIUM IODIDE AS A THYROID-BLOCKING AGENT IN A RADIATION EMERGENCY

Request for Submissions of New Drug Applications and Notice of Availability of Labeling Guidelines

AGENCY: Food and Drug Administration.

ACTION: Notice.

SUMMARY: The Food and Drug Administration (FDA) requests submissions of new drug applications (NDA's) for potassium iodide in oral dosage forms for use as a thyroid-blocking agent in a radiation emergency. The approval of oral dosage forms of potassium iodide as a thyroid-blocking agent for use in a radiation emergency would be one step in meeting the responsibilities of the Department of Health, Education, and Welfare (DHEW) to State and local governments for radiological emergency response planning. The agency encourages interested persons to submit NDA's in the interest of the public safety. The agency is also announcing the availability of labeling guidelines for potassium iodide for such use.

SUPPLEMENTARY INFORMATION: By FEDERAL REGISTER notice of December 24, 1975 (40 FR 59494), the General Services Administration (GSA) outlined the responsibilities of several Federal agencies concerning certain emergency response planning guidance that the agencies should provide to State and local authorities. The Department of Health, Education, and Welfare (DHEW) is responsible for assisting State and local authorities in developing plans for preventing adverse effects from exposure to radiation in the event that radioactivity is released into the environment. These plans are to include the prophylactic use of drugs that would reduce the radiation dose to specific organs due to the sudden release into the environment of large quantities of radioactivity that might include several radioactive isotopes of iodine.

BACKGROUND

The GSA notice of December 24, 1975, concluded that there is an exceedingly low probability that incidents will occur involving either the use of radioactive materials in fixed nuclear facilities or the transportation of those materials. Because of the possible increase in number of nuclear power plants, however, several Federal agencies are identifying those possibilities, however remote, that could adversely affect the public, should an incident occur. One possibility is the sudden release of large quantities of radionuclides, which might include a number of isotopes of radioiodine, into the environment. When radioiodines are inhaled or ingested, they rapidly accumulate in the thyroid gland and are metabolized into organic iodine compounds. These compounds could reside in the thyroid gland long enough to allow for local radiation damage, resulting in thyroiditis, hypothyroidism, or thyroid neoplasia with either benign or malignant characteristics. Therefore, it is considered in the public interest that State and local authorities be prepared to take effective measures to prevent or curtail markedly the accumulation of radioiodines by the thyroid gland, should such an incident occur. These measures may include the use of a thyroid-blocking agent.

An ad hoc committee to the National Council on Radiation Protection and Measurements (NCRP), which included FDA representatives as consultants, studied the feasibility of using certain drug products as thyroid-blocking agents to reduce radiation dose to the thyroid gland. The NCRP, located in Bethesda, Maryland, is a nonprofit corporation chartered by Congress in 1964 to collect, analyze, develop, and disseminate information and recommendations about radiation protection. The NCRP is made up of 56 scientific committees, composed of experts having detailed knowledge and competence in the particular area of the committee's interest. An NCRP report published August 1, 1977 (NCRP Report No. 55, "Protection of the Thyroid Gland in the Event of Release of Radioiodine") discusses the safety and efficacy of thyroid-blocking agents and recommends that potassium iodide be considered for thyroid-blocking purposes under certain emergency conditions.

The report discusses stockpiling thyroid-blocking agents at appropriate outlets for ease of distribution in the event their use is necessary in a radiation emergency.

But there are hundreds of radioactive poisons which would be let loose in a "radiation emergency" besides radioactive iodine.

And there's no magic pill for the horror.

CHAPTER FOUR

Medical Consequences

Cancer. Who reading this doesn't know someone who has recently died—or is dying right now—from this modern day scourge? And why?

Cancer was the number eight cause of death in America in 1900, accounting for four per cent of all deaths caused. Today (after heart disease) it is the second leading cause of death—accounting for twenty per cent of all deaths. And the cancer rate continues to climb.

It's a horrible way to die. Wild, uncontrolled growth starts off in the body, spreading, stopping for a while, then spreading again. A piece of lung is cut out, a piece of stomach, and the cancer goes into "remission" for a bit, then spreads again. Cancer. The life process out of control.

All major organizations involved in studying cancer—the World Health Organization of the United Nations, the National Cancer Institute, and the journals specializing in cancer—agree: between seventy and ninety per cent of cancer is the result of environmental factors, particularly the contaminants man has brought in recent decades to the environment.

Radioactivity. Cigarette smoke. Pesticides. Plastics. Solvents. Air pollutants. Contaminants in our air, water and food. Cancer researchers try to break down the responsibility: this much from radioactivity, this much from smoking, this much from petro-chemicals, and so on. Which contaminant does what damage is unclear. What is undisputed is that taken together humanity has visited a plague onto itself.

"If one thousand people died every day of cholera, swine flu, or food poisoning, an epidemic of major proportions would be at hand and the entire country would mobilize against it," declares Dr. Samuel S. Epstein in *The Politics of Cancer.* "Yet cancer claims that many lives daily, often in prolonged and agonizing pain, and most people believe they can do nothing about it . . . . Cancer, they think strikes where it will, with no apparent cause . . . . But cancer has distinct, identifiable causes," he explains. "Cancer is caused mainly by exposure to chemical or physical agents in the environment."

"There is a great misunderstanding by the public about the nature and causes of cancer, a disease which has now reached epidemic proportions in the United States. Cancer is not caused by some inexplicable miasma, although it may seem that way to the uninformed. Instead, most scientists now agree that the overwhelming majority of cancers are environmentally caused," notes *Malignant Neglect.* "As such they are largely preventable."

However, to *prevent* cancer means to eliminate those contaminants in society which cause it.

There are those with huge vested interests in the products and processes which cause cancer. So instead of our working on ways to prevent cancer, we are channeled into a preoccupation with a cancer cure. But for cancer, the only cure is prevention. Once life is sent out of control by cancer, there can be some holding off, if one is lucky, but for most, death soon arrives. Prevention is life's only salvation.

We are steeping ourselves in a sea of radioactivity. Everywhere in the nuclear cycle radioactivity is produced and once created, stays on and on, ending up in the water, in the earth, in the air and quite likely sooner or later, in you or someone you love.

And a massive dose, even a mid-range dose of radioactivity, the kind you'd get from a nuclear plant accident, is not necessary to produce cancer. "Routine" radioactive emissions will do it.

Radioactivity causes cancer and kills by altering the natural balance of cells, what is called their "electric potential." Struck by radioactivity, cells may die quickly or lose their ability to duplicate, or they produce abnormal, cancerous cells which become cancerous growths. And injured cells pass on genetic deformities. Radioactivity, even the tiniest amount, can strike the control mechanism within cells to send them out of control.

It takes but one affected cell, which goes on to reproduce, to trigger cancer.

The way radioactivity affects life has been sometimes compared to what would happen if you removed the back of a television set and cut one wire someplace—any wire. You can be sure that it will not improve the way the TV set works. And it can easily lead to complications, that broken circuit affecting another circuit, spreading malfunction. Likewise, when radioactivity clips apart one of the billions of circuits of life, the result can be a black picture.

From the outset of man's making radioactivity, it became clear that cancer was a constant by-product. Marie Curie and her daughter, Irene, well known for their pioneer work with X-rays, both died of leukemia (blood cancer). Many early radiologists died from overexposure to radiation.

But it was not until the past several decades, with vast quantities of radioactivity entering our environment from civilian and military uses of nuclear power, that the radioactivity/cancer connection became manifest on a wide scale.

It was hard to predict what would happen as radiation production increased, because low-level radiation has an "incubation" or "latency" period of five to forty years. Only in recent years are we beginning to see that doses of radioactivity once thought to be "safe" are *not safe*—indeed, that there is no safe dose of radiation. This is the truth. And as research has confirmed it, the nuclear establishment has made strenuous efforts to suppress this fact.

Scientific report after scientific report shows that the soldiers who viewed atomic bomb tests in the 1950's are now falling from cancer, that workers at shipyards handling nuclear vessels are getting cancer, that workers at nuclear laboratories and fabricating centers get cancer at a record rate, that people who live near nuclear plants are developing cancer at a far higher rate than those who do not. The reason: they have been absorbing radiation at rates which had not before been thought dangerous.

Drs. John Gofman and Arthur Tamplin* have projected that if the average radiation exposure of the U.S. population is to reach what the U.S. government allows as the limit—.17 rems per year—there will "be an excess of 32,000 cases of fatal cancer plus leukemia per year, and this would occur year after year."

The Advisory Committee on the Biological Effects of Ionizing Radiation of the National Academy of Sciences now estimates that one-half of one per cent of America's population will develop cancer from man-made sources of radiation at some time during their lifetime.** That's a million cancers.

And still, because of the delay factor for cancer from low-level radiation, we have not fully grasped the magnitude of the problem—that with nuclear power we are planting a harvest of cancer deaths.

But what about "background" radiation?

There have always been some natural sources of radiation in the world—from cosmic rays and uranium and the several other radioac-

**Poisoned Power*, Rodale Press, Inc., Emmaus, Pa. 1971.
**Report of BEIR 3 Committee, May, 1979.

tive elements in the earth's crust–but that doesn't make it safe, either. Where radiation is highly concentrated in the earth's ground (areas such as Kerala, India), studies have shown corresponding effects on health. Drs. Gofman and Tamplin have calculated that background radiation in the U.S.–some .1 rem or 100 millirem a year–causes 19,000 cancer and leukemia deaths yearly and as many as 588,000 deaths resulting from genetic defects which produce heart disease and diabetes. But nothing can be done about background radiation. It is something we have to live with and minimize, if possible. There is no reason to add to it.

As Dr. Helen Caldicott, president of Physicians for Social Responsibility, says: "As a physician I contend that nuclear technology threatens life on our planet with extinction. If present trends continue, the air we breathe, the food we eat, and the water we drink will soon be contaminated with enough radioactive pollutants to pose a potential health hazard far greater than any plague humanity has ever experienced. Unknowingly exposed to these radioactive poisons, some of us may be developing cancer right now. Others may be passing damaged genes, the basic chemical units which transmit hereditary characteristics, to future generations. And more of us will inevitably be affected unless we bring about a drastic reversal of our government's pro-nuclear policies."

How does radioactivity kill?

Radioactivity, explains Drs. Gofman and Tamplin in their key book on the subject, *Poisoned Power,* causes "massive, non-specific disorganization or injury of biological cells and tissues. Biological cells are remarkably organized accumulations of chemical substances, arranged into myriad types of sub-structural entities within the cells. The beauty of such organization can only be marveled at when revealed under the high magnifications of such instruments as the electron microscope or the electron scanning microscope. In stark contrast, there is hardly anything specific or orderly about the ripping of chemical bonds or of electrons out of atoms. Rather, this represents disorganization and disruption. Perhaps a reasonable analogy would be the effect of jagged pieces of shrapnel passing through tissues."

From a massive dose of radioactivity, the kind that would be released by a nuclear plant accident, there is "a general disruption of the charge within the brain cells, causing swelling and hemorrhage within the brain," explains Harvard University-educated physician and surgeon Dr. Stephen Sigler. There is "acute radiation sickness, bloody diarrhea, vomiting, aplastic anemia, skin burns" from a dose of two or three thousand rem, and death will follow in a matter of hours. "There is cerebral edema," explains Dr. Sigler. "The brain is so disrupted that it swells and death ensues because of swelling and compression of the vital structures. Another form of acute radiation sickness is sudden cessation of multiplication of gastro-intestinal cells. Massive gastro-intestinal bleeding follows. A third form is acute aplastic anemia where there are no clotting factors or white cells or red cells because of sudden cessation of bone marrow and lymph node production."

The U.S. Atomic Energy Commission's "Operational Accidents and Radiation Exposure Experience" gives this example of a high level radioactivity death:

> FATAL INJURY ACCOMPANIES CRITICALITY ACCIDENT
> Los Alamos, N. Mex., Dec. 30, 1958
>
> The chemical operator introduced what was believed to be a dilute plutonium solution from one tank into another known to contain more plutonium in emulsion. Solids containing plutonium were probably washed from the bottom of the first tank with nitric acid and the resultant mixture of nitric acid and plutonium-bearing solids was added to the tank containing the emulsion. A criticality excursion occurred immediately after starting the motor to a propeller type stirrer at the bottom of the second tank.
>
> The operator fell from the low stepladder on which he was standing and stumbled out of the door into the snow. A second chemical operator in an adjoining room had seen a flash, which probably resulted from a short circuit when the motor to the stirrer started, and went to the man's assistance. The accident victim mumbled he felt as though he was burning up. Because of this, it was assumed that there had been a chemical accident with a probable acid or plutonium exposure. There was no realization that a criticality accident had occurred for a number of minutes. The

quantity of plutonium which actually was present in the tank was about ten times more than was supposed to be there at any time during the procedure.

The employee died 35 hours later from the effects of a radiation exposure with an *average* whole-body dose calculated to be about 5,000 rem.

Two other employees received radiation exposures of 134 and 53 rem, respectively. Property damage was negligible. (*See* TID-5360, Suppl. 2, p. 30; *USAEC Serious Accidents* Issue #143, 1-22-59.)*

When the victim said he was "burning up" his skin was indeed cherry red. In minutes he was vomiting and discharging a profuse diarrhea. By the time he got to the hospital he was in shock. Mercifully, he was soon dead.

The bone marrow, where new blood cells must be made continually and rapidly (a red blood cell normally lives 120 days) is the central target for the mid-range dose of radioactivity, about 500 rem, also what a nuclear plant accident would bring.

Aplastic anemia is the failure of the bone marrow to produce blood platelets, the clotting factors that allow blood to coagulate and keep us from bleeding to death. Leukemia is a blood cancer in which abnormal blood cells are produced. Radioactivity "by its alteration of the bio-electric potentials within the cells of the bone marrow," explains Dr. Sigler, can cause aplastic anemia or leukemia. In these cases death is somewhat slower than from a massive radiation dose.

The British Medical Research Council describes the consequences of mid-range radioactivity this way:

"The first effect . . . is a sensation of nausea developing suddenly and soon followed by vomiting and sometimes by diarrhea. In some people, these symptoms develop within half an hour of exposure; in others, they may not appear for several hours. Usually, they disappear after two or three days. In a small proportion of cases, however, the symptoms persist; vomiting and diarrhea increase in intensity; exhaustion, fever, and perhaps delirium follow; and death may occur a week

*From: "Operational Accidents and Radiation Exposure Experience," U.S. Atomic Energy Commission, April, 1965.

or so after exposure. Those who recover from the phase of sickness and diarrhea may feel fairly well, although examination of the blood will reveal a fall in the number of white cells. Between the second and fourth weeks, however, a new series of ailments, preceded by gradually increasing malaise, will appear in some of those exposed. The first sign of these developments is likely to be partial or complete loss of hair. Then, from about the third week onwards, small hemorrhages will be noticed in the skin and in the mucous membranes of the mouth, which will be associated with a tendency to bruise easily and to bleed from the gums. At the same time, ulcerations will develop in the mouth and throat, and similar ulceration occurring in the bowels will cause a renewal of the diarrhea. Soon the patient will be gravely ill, with complete loss of appetite, loss of weight, and sustained high fever. Feeding by mouth will become impossible, and healing wounds will break down and become infected. At this stage the number of red cells in the blood is below normal, and this anemia will increase progressively until the fourth or fifth week after exposure. The fall in the number of white blood cells, noted during the first two days after exposure, will have progressed during the intervening symptomless period, and will by now be reaching its full extent. The changes in the blood count seriously impair the ability to combat infection, and evidence from Nagasaki and Hiroshima shows that infections of all kinds were rife among the victims of the bomb. Many of those affected die at this stage and, in those who survive, recovery may be slow and convalescence prolonged; even when recovery appears to be established, death may occur suddenly from an infection which in a healthy person would have only trivial results."

The consequences of low-level radioactivity are yet more delayed—with cancer showing up five to forty years later, after an incubation or latency period.

Why the delay?

Explains Dr. Sigler: "It takes a while for the genetic effects of these altered electric potentials produced by the radiation to accumulate in enough cells to actually produce the abnormal clinical picture of leukemia. In other words, out of a million duplicating marrow cells one or two might be affected by the radiation. It would take a while for this clone to multiply enough to actually get to a clinical level. The latency period is the time required for the atypical cells to become predominant in the cell population, for the multiplication process to get large enough to dominate the blood picture or to emerge as a tumor in an organ."

Often the pathway for low-level radioactivity is ingestion—through the food we eat, the water we drink or the air we breathe.

Life is defenseless against the 200 fission products created in fission—the lethal twins of safe, stable elements in nature. The body cannot tell safe strontium from strontium-90, its lethal radioisotope, or safe cesium from cesium-137, its lethal radioisotope. Life cannot distinguish between the stable chemicals it needs and the radioactive twins that will harm it.

And many of the radioactive twins are drawn to specific organs of the body—the organs that require their safe, stable counterparts.

The radioactive twins concentrate in whatever organ they're attracted to. Many go to the bones and bone marrow. Others go to the soft tissue. That's "the bowel, or the lung, or the kidney," explains Dr. Sigler, and "wherever it concentrates" tumors may well appear.

Explain Drs. Gofman and Tamplin: "We must recall that the radioactive elements produced in a nuclear reactor behave almost precisely as do their non-radioactive counterparts. For example, radioactive iodine-131 behaves chemically and biologically just as does stable, or non-radioactive, iodine . . . . The thyroid gland has a special affinity for iodine. As a result, the thyroid accumulates far, far more iodine from an ingested dose than any other body organ does. The thyroid uses iodine to manufacture its major active hormone, thyroxin. The radioactive forms of iodine (iodine-131 is one) behave just as non-radioactive iodine would when taken in with food, accumulating preferentially in the thyroid gland. As a result, that tissue receives a far higher radiation dosage in rems from the decaying radioactive iodine than other tissues of the body do. Naturally, radioactive iodine has its major biological effects on the thyroid gland, compared with its effect on other cells in the whole body."

Here's a chart showing which organs some of the radioisotopes attack:

IONIZING RADIATION

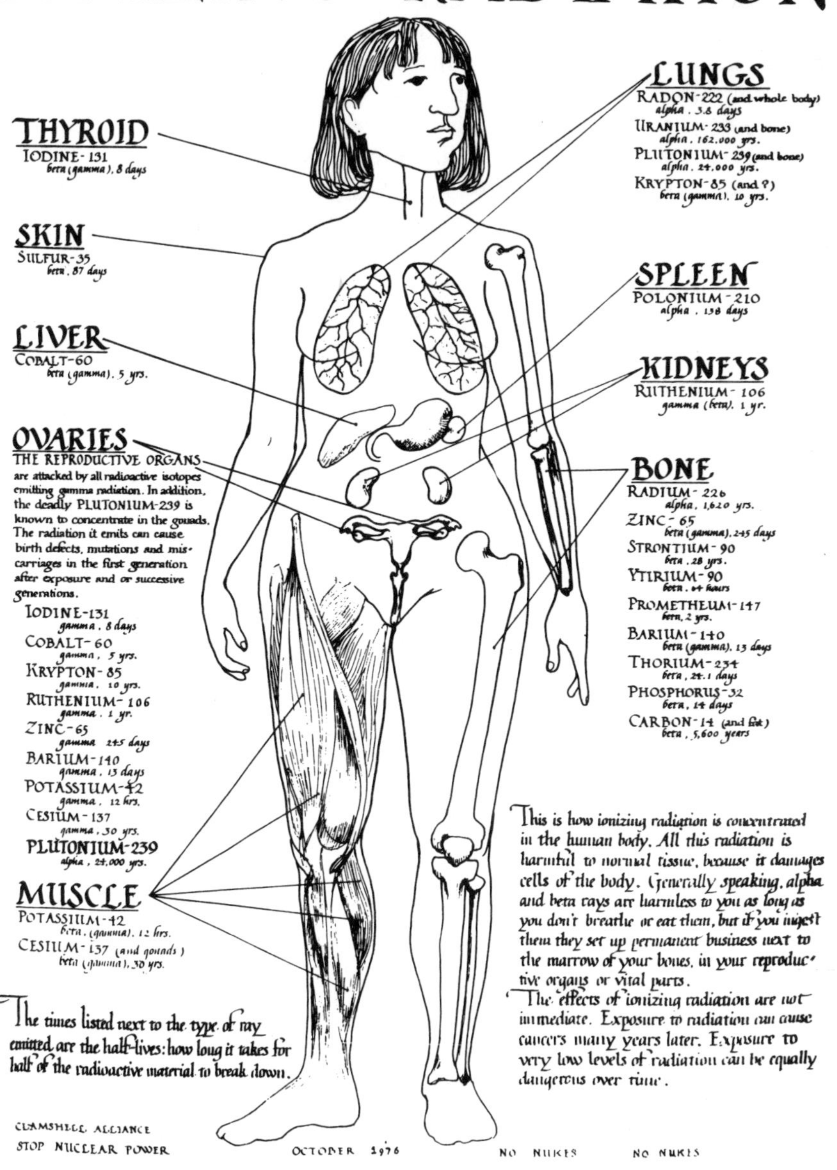

What about the effects of plutonium?

As Nobel Laureate in medicine, James D. Watson has declared: "I fear that when the history of this century is written, that the greatest debacle of our nation will be seen not to be our tragic involvement in Southeast Asia but our creation of vast armadas of plutonium, whose safe containment will represent a major precondition for human survival, not for a few decades or hundreds of years, but for thousands of years more than human civilization has so far existed."

Explains Dr. Caldicott: "We're talking about a substance that is so incredibly toxic that everybody who comes in contact with it and gets it into their lungs will die of a lung cancer. You don't know you've breathed it into your lungs. You can't smell it, you can't taste it, and you can't see it. Nor can I, as a doctor, determine that you've got plutonium in your lungs. When a cancer develops, I can't say that cancer was made by plutonium. It doesn't have a little flag saying, 'Hey, I was made by plutonium.' And you'll feel healthy for 15 to 20 to 30 years while you're carrying around that plutonium in your lung, till one day you get a lung cancer. It's a very insidious thing. We have to teach people that it takes a long time to get the cancer. If I die of a lung cancer produced by plutonium, and I'm cremated, the smoke goes out the chimney with the plutonium, to be breathed into somebody else's lungs–ad infinitum for half a million years."

Plutonium, she goes on, is also absorbed from the lungs "into the blood stream where it is carried to the liver, to produce a very malignant liver cancer, to bone, where like strontium-90 it causes osteogenic sarcoma and leukemia, and it is selectively taken up from circulation by the testes and ovaries where, because of its incredible gene changing properties, it may cause an increased incidence of deformed and diseased babies, both now and in future generations. Plutonium also crosses the placenta, from the mother's blood into the blood of the fetus, where it may kill a cell responsible for development of part of an organ, e.g. heart, brain, etc., causing gross deformities to occur in the developing fetus. This mechanism for production of fetal deformities is called teratogenesis and is different from the deformities caused by genetic mutation in the egg or sperm because although the basic gene structure of the cells of the fetus is normal, an important cell in the developing fetus has been killed leading to a localized deformity, similar to the action of the drug thalidomide."

Says Dr. Caldicott: "Nuclear power poses the greatest public health hazard the world has ever encountered because of the inevitable contamination of the biosphere with plutonium and radioactive wastes.

Cessation of all forms of nuclear power is the ultimate form of preventive medicine."

What about genetic damage?

At all dose levels radioactivity injures genes. The only reason it is not of prime concern in those people who suffer high and mid-range radiation doses is that they are not likely to go on and reproduce—indeed not likely to survive at all.

Drs. Gofman and Tamplin explain genetic damage. "Genes are the units of information within the chromosome. They are composed of the chemical known popularly as DNA (deoxyribo-nucleic acid). Radiation can produce a chemical alteration in a part of a single gene, so that the gene functions abnormally thereafter, providing the cell with false direction. When such cells divide, the altered gene may be reproduced in the descendant cells."

Once, scientists thought that under a given "threshold" of radiation, life would not be damaged.

But even the U.S. government eventually was admitting it just was not so. A 1966 Atomic Energy Commission pamphlet by Isaac Asimov (a main AEC nuclear power writer) and Theodosius Dobzhansky, entitled "The Genetic Effects of Radiation," concedes:

> It is generally believed that the straight line continues all the way down without deviation to very low radiation absorptions. This means there is no "threshold" for the mutational effect of radiation. No matter how small a dosage of radiation the gonads receive, this will be reflected in a proportionately increased likelihood of mutated sex cells with effects that will show up in succeeding generations.
>
> Suppose only one sex cell out of a million is damaged. If so, a damaged sex cell will, on the average, take part in one out of every million fertilizations. And when it is used, it will not matter that there are 999,999 perfectly good sex cells that might have been used—it was the damaged cell that *was* used. That is why there is no threshold in the genetic effect of radiation and why there is no "safe" amount of radiations insofar as genetic effects are concerned. However small the quantity of radiation absorbed, mankind must be prepared to pay the price in a corresponding increase of the genetic load.

Nobel Award-winning geneticist Dr. Joshua Lederberg has warned that present radiation standards allow for a ten per cent increase in the mutation rate—which could spell disaster, he feels, for the human species. Said the AEC in a 1968 report: "A 10% increase in mutation rate, whatever it might mean in personal suffering and public expense, is not likely to threaten the human race with extinction . . . . [It] is bearable if we can convince ourselves that the alternatives of abandoning radiation technology altogether will cause still greater suffering. If the number of those affected is increased, there would come a crucial point, or threshold, where the slack could no longer be taken up (by those not affected). The genetic load might increase to the point where the species as a whole would generate and fade toward extinction—a sort of 'racial radiation sickness.' "*

What about the government's allowable levels of radiation exposure?

Say Drs. Gofman and Tamplin, "all the evidence, both from experimental animals and from humans" shows "that even the smallest quantities of ionizing radiation produce harm, both to this generation of humans and future generations. Furthermore, it appears that progressively greater harm accrues in direct proportion to the amount of radiation received by the various body tissues and organs . . . . Nuclear electricity generation has been developed under the false illusion that there exists some safe amount of radiation . . . . Obviously any engineering development proceeding under an illusion of a wide margin of safety is fraught with serious danger. What is more, the false illusion of a safe amount of radiation has pervaded all the highest circles concerned with the development and promotion of nuclear electric power. The Congress, the nuclear manufacturing industry, and the electric utility industry have all been led to believe that some safe amount of radiation does indeed exist. They were hoping to develop this industry with exposures below this limit—a limit we now know is anything but safe."

They ask: "How, under such circumstances, is it even conceivable that so many important industrial and governmental leaders were so totally and seriously misled, misled to the point of launching a multibillion dollar industry based upon a dangerously false premise?"

They point to the Atomic Energy Commission, "with the impossible

*"The Genetic Effects of Radiation."

dual role of promoter of atomic energy and protector of the public from radiation" having "historically suffered from false optimism."

And it took a while—that five to forty year latency period—before the widespread effects of radiation became obvious. Still, those involved in the nuclear establishment put their wishful thinking and self-interest ahead of the evidence. They still do.

For instance, this government pamphlet—still in use—for workers at Brookhaven National Laboratory, in large measure a government-funded research and development facility for nuclear power, explains how people "can learn to live with radiation."

A WORD TO THE STAFF—

This pamphlet on The ABC's of Radiation has been prepared primarily for those of you at Brookhaven who are not already familiar with the subject. In it you will find information about radiation and the precautions which the Laboratory takes to insure your safety and that of this whole area.

Everyone on the staff should know what radiation guidelines are, and should be familiar with our routine protective measures, so that he can cooperate with our radiation protection program and explain it to his friends and neighbors.

Everyone here is associated with radiation in one way or another. It should be regarded with respect, but it need not be feared. Safety is attainable if the necessary rules and procedures are followed. Danger lurks only for the uninformed or careless.

G. H. Vineyard

GEORGE H. VINEYARD

Foreword

It is said that we are living in the Atomic Age. It should be a period of great advancement. As citizens we should add to our knowledge of what Atomic Energy means. A good place to start is with the simple facts of Radiation.

IS RADIATION DANGEROUS TO YOU?

It can be; but it need not be.

Danger from radiation depends upon the degree of exposure. How dangerous is fire, or exposure to the sun? How dangerous is electricity? It depends

upon your exposure. We all use fire. We use electricity, but we do not take chances. We have learned to live with these agents, and we can learn to live with radiation too.

Radiation effects on people were noticed shortly after the discovery of x-rays. These effects resulted from extreme exposures due to ignorance. We now have special instruments to detect and measure all types of radiation. The "rem" has been adopted as the basic unit for radiation protection. It is simply a label for a certain amount of radiation, just as the word "inch" is the label or word used to describe a certain distance.

With this unit of measurement, we are able to compare radiation exposure with its effects on living tissue. Years of experience with x-rays and radium and thousands of experiments with animals have made it possible to judge how much radiation we can receive without producing observable harmful effects. This level is considerably higher than the amount of exposure which Brookhaven staff members are permitted to receive. The Laboratory's standard operating limit is a continuing exposure of no more than 3 rem per quarter year. In addition, the total exposure received at work is not allowed to be more than 5 rem times the number of years since the person was 18 years old.

How much radiation can you stand? The important thing is that you do not take too much at one time. Small exposures with intervals in between can add up to a fairly high amount without harmful effects because cells either recover by themselves or can be replaced by other cells. Furthermore, you may safely expose a portion of the body to a much higher amount than is permissible for the entire body.

Radioactive materials can be harmful if within or on the body. You should, therefore, avoid inhaling radioactive substances or getting them into your food or drink, just as you would avoid taking in arsenic, lead, or other poisonous substances. The amount might be insignificant, but there is no sense in taking chances.

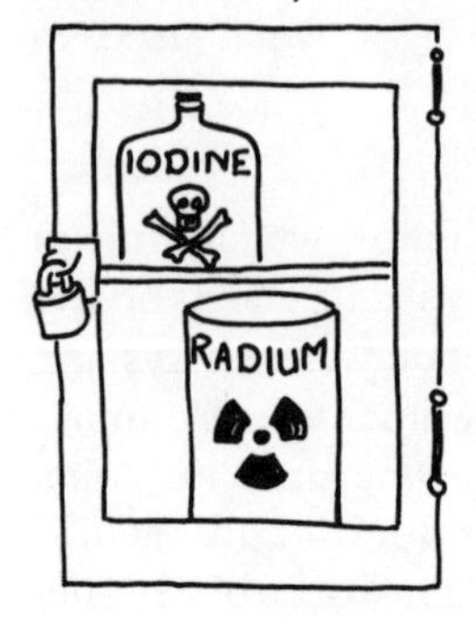

Radioactive materials differ widely in the rate at which they lose radioactivity. The length of time they are kept in the body also varies. Radium and plutonium remain active for thousands of years and may be retained for long periods in the body, while such elements as radiosodium will not only be quickly eliminated but decays in a few days. Naturally, you must be careful to avoid taking even small amounts of the more poisonous materials into your mouth or lungs. This is why eating or smoking is forbidden in most radiation areas.

When you have a chest x-ray taken with a common type of automatic equipment, you receive approximately one-tenth rem. During an examination of the stomach or intestines, patients frequently receive a series of exposures over a period of a few hours which may total as high as 15 or 20 rem. To render a person permanently sterile, the sex organs alone would have to receive a single exposure of 400 to 800 rem, and even more if the total amount were not given at one time. Between 300 and 500 rem of x- or gamma radiation given to the whole body in one, short dose could prove fatal without medical attention. This, however is an enormous amount, more than 100 times greater than Brookhaven's operating limit for any single exposure. By the same token, a person could receive a single exposure 10 times greater than the Laboratory's operating limit, and incur virtually no observable harmful effects. However, a word of caution is in order here. Years of study with many chemical poisons have shown that these materials are harmful above a certain dose, or threshold; and that below that dose they have been found to produce essentially no harmful effects. Carbon monoxide is a good example. We know it is a potent poison, yet we breathe small amounts of it constantly. In spite of these observations, however, we try to minimize our exposure to such materials. This philosophy is also true in dealing with radiation. In other words, radiation scientists have not been able to *observe* harmful effects at very low radiation doses, but they cannot guarantee that *some* harm hasn't been done. To be on the safe side then we must assume that there is some risk involved in receiving *any* radiation, and we try to balance this risk with the benefits received for taking it. This concept of weighing benefits versus risk is not new. For example, we make the same type of evaluation whenever we drive a car and accept the risk of being involved in an accident.

RADIATION AREAS

As a general precaution, sources of radiation are confined to special "controlled areas" in which they are either roped off or are clearly indicated. These controlled areas fall into two defined categories:

"Radiation Areas" – where a person could receive a dose of more than 5 millirem in one hour, or more than 100 millirem in a 5 day work week, and

"High Radiation Areas" – where a person could receive a dose in excess of 100 millirem in one hour. All radiation areas are marked with signs bearing the purple radiation symbol on a yellow background, and indicating the nature of the radiation source. Near the source itself will be a warning sign with the purple symbol on a yellow background and a card stating the type of radiation, its strength, and the precautions to be taken.

HOW TO LIMIT EXPOSURE TO RADIATION

Exposure to radiation can be limited in three ways: 1) in time, 2) by distance, and 3) by shielding.

If you must work near radiation, the simplest way to limit your exposure is to stay in the vicinity as short a time as possible. If there is a time limit on your job, observe it.

A second method is to maintain a safe distance between you and the source of radiation. If in doubt as to what distance is safe, consult your supervisor or the Health Physicist. In general, the effect of radiation falls off sharply as you increase your distance from its source. Double the distance and your exposure is cut to one quarter.

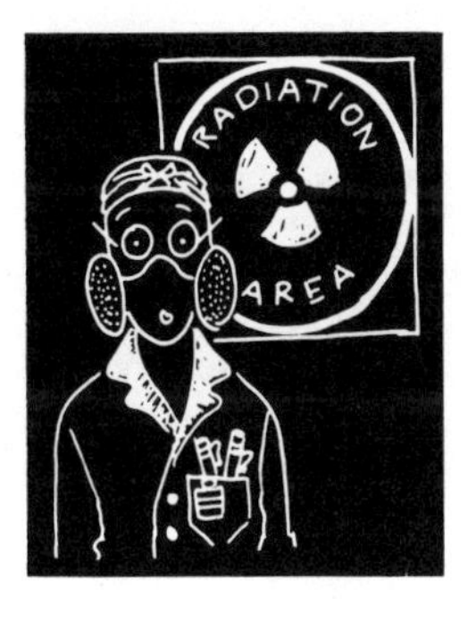

To guard against contamination, special protective clothing is available in selected controlled areas. Its use protects the wearer and helps to confine radioactive contamination within these areas. Laboratory coats or coveralls are widely used; in some locations caps, shoe covers, canvas or rubber gloves, masks or respirators are also used. Protective clothing worn where radioactive materials are present is specially marked and washed.

Alpha-, beta-, and gamma-rays are not "catching" like a cold. Unless your hands, feet, or clothing are actually contaminated with materials

which give off radiation, you cannot take these rays home to your family. If you wear protective clothing, wash with soap, and check your hands and feet with instruments, you are perfectly safe.

Scientists have been blowing the whistle as the years have gone by and the effects of radioactivity on people became apparent.

A pivotal study was made in 1958 by Dr. Alice Stewart of Oxford University.

A SURVEY OF CHILDHOOD MALIGNANCIES

The present survey is based on an earlier study of the vital statistics relating to leukaemia (Hewitt, 1955). This had revealed an unusual peak of mortality in the third and fourth years of life which indicated that the subsequent survey should, in the first instance, be restricted to children. The earlier investigation had also led to the suggestion that it might be particularly worth while to study modern innovations, such as radiology.

An attempt was made to trace all children in England and Wales who had died of leukaemia or cancer before their tenth birthday during the years 1953 to 1955 (case group) and to compare their pre-natal and post-natal experiences with those of healthy children (control group).

Available Cases.—The total number of deaths in the category required was 1,694, of which 792 were ascribed to leukaemia and 902 to other cancers

---

The pre-natal and post-natal experiences of a large group of children who recently died of malignant diseases have been compared, point by point, with the experiences of a similar group of live children.*

Dr. Stewart concluded that children whose mothers received very small amounts of radiation through X-rays during pregnancy, amounts not thought dangerous, had twice the risk of developing leukemia before the age of ten than those whose mothers had not. Subsequent studies have confirmed her findings.

The nuclear establishment need not be surprised. The AEC publication, "Your Body and Radiation," shows radiation-induced deformities:

Figure 21 *A shows a normal chick embryo 10 days after fertilization. C is a 10-day chick that had been irradiated with cobalt-60 gamma rays on the sixth day after fertilization. Note deformities of beak and toes and generalized hemorrhage and swelling. B shows a normal chick embryo 13 days after fertilization. D is a 13-day chick that had been irradiated on the sixth day. In addition to the defects seen in C, there is serious growth retardation.*

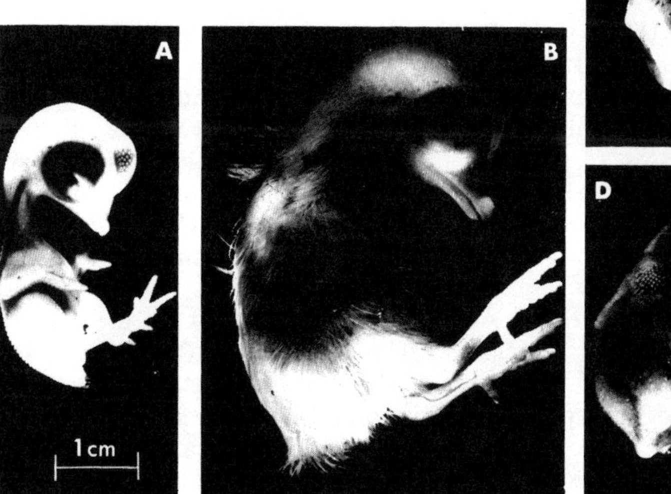

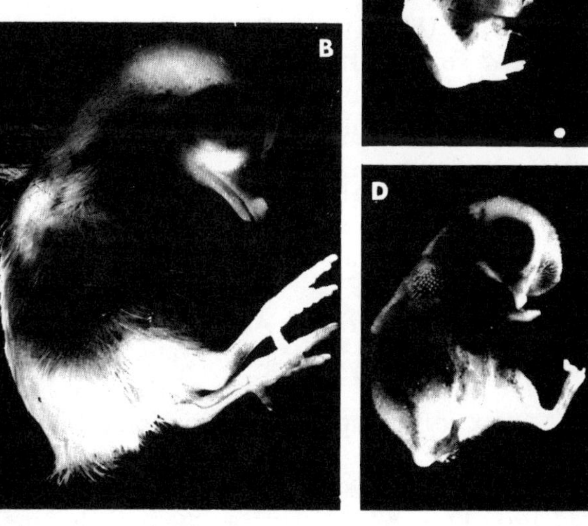

*From the *British Medical Journal,* 1:1495, 1958.

Still its approach was "play it down."

Congressman Chet Holifield (who later became chairman of the Congressional Joint Committee on Atomic Energy) declared in the 1950's:

"I believe from our hearings that the AEC approach to the hazards from bomb test fallout seems to add up to a party line—'play it down'. As custodian of official information, the AEC has an urgent responsibility to communicate the facts to the public. Yet time after time there has been a long delay in issuance of the facts, and often times the facts have to be dragged out of the agency by the Congress. Certainly it took our investigation to enable some of the Commission's own experts to break through the party line on fallout."

Study after study has shown the direct link between radioactivity in any amount and death and cancer. The nuclear establishment ignores these studies, still tries to play them down and discredit the whistle-blowers.

In the 1960's, Dr. Ernest Sternglass, professor of radiological physics at the University of Pittsburgh, made numerous studies and found correlations between low-level radiation and mortality and cancer. In response, the AEC assigned Drs. Gofman and Tamplin, top scientists at the AEC's Livermore Radiation Laboratory, to review and, the AEC hoped, to discredit Dr. Sternglass' work.

They, in turn, came away after extensive study with the conclusion that radioactivity is a far greater threat to life than had been thought. "We were assigned to evaluate the hazards of atomic radiation by the U.S. Atomic Energy Commission in 1963," they write in *Poisoned Power.* "It was our job to assess the cost in human disease and death for all sorts of proposed and on-going nuclear energy programs, including nuclear electricity . . . . Our work showed that previous estimates . . . were ten to twenty times too low. The new evidence, on radiation-induced human cancer-plus-leukemia, from Japan, from Great Britain and from Nova Scotia, were now all telling us one story—radiation is a greater factor in cancer-leukemia than had been previously realized."

In 1972, the National Academy of Sciences set up the Advisory Committee on the Biological Effects of Ionizing Radiation to look into the findings of Drs. Gofman and Tamplin that the AEC's "allowable exposure" standards for radiation would mean 32,000 additional cases of fatal cancer and leukemia yearly in America.

The committee, supporting their findings, declared that the AEC radiation limit would cause a "most likely" increased annual number of deaths by 5,000 to 7,000 persons, possibly as many as 15,000. And these figures did not take into account the effect of radiation on fetuses. But the AEC moved against Drs. Gofman and Tamplin, forcing them to resign from Livermore.

Again and again the government has dealt with such studies by punishing their authors.

Dr. Thomas Mancuso of the University of Pittsburgh was assigned by the AEC to study the "biological effects, if any, of low level ionizing radiation among workers employed in atomic energy facilities" and to determine why 3,520 people who worked at the government's Hanford nuclear facility between 1944 and 1972 had died. Dr. Mancuso found the cancer mortality rate among Hanford workers, workers who had received amounts of radiation well within the acceptable level, substantially higher than that of the general population.

Dr. Mancuso subsequently had his funding cut, and the government moved to transfer all the data in the research project to a facility under its control at Oak Ridge National Laboratory.

Dr. Mancuso stated that such arbitrary transfer undermined impartial scientific inquiry:

> A research project can be administratively terminated or transferred, but in so doing, there is no way to transfer the insights of the investigators nor experience and caution gained through many years in the development of the data, knowledge of the specific areas which must be considered as well as the particcular promising avenues of analyses. This principle applies to every long-term epidemiological scientific investigation that covers a span of many years.
>
> This decision to terminate the project at the University of Pittsburgh, in light of the positive findings of a definite relationship between work exposure to ionizing radiation and cancer is, in my opinion, not in the best interest of science.*

Another instance: Dr. Irwin D. J. Bross of the Roswell Park Memorial Institute, a cancer research facility run by the New York State Department of Health, lost his funding from the National Cancer In-

*From "Study of the Lifetime Health and Mortality Experience of Employees of ERDA Contractors," Thomas F. Mancuso, M.D., Final Report, July 13, 1977.

stitute under similar circumstances. His study demonstrated the correlation of cancer with low-level radiation through X-rays in a group of 13 million people. Again radiation levels far lower than expected to cause damage were leading to cancer.

"What happened to me is only important as a clearcut example of how federal science agencies operate," Dr. Bross later explained. "What really matters is the effect of 'Big Science' practices on science and on the public health. The cut-off of funding denies the public the truth about the hazards of low-level radiation which the public has every right to know."

Dr. Bross notes: "A generation ago some persons considered 1,000 rems to be perfectly safe because, while there might be some immediate ill effects, the persons exposed didn't drop dead on the spot. Since then, the so-called safe level has dropped and dropped and dropped. For a while it was 500, then 200, then 100, then 50, then 20, 15, now 5 (for nuclear workers) and hopefully, in the near future, one-half rem. What is most striking about the shrinking margin of safety is that, wherever the level was, the health physicists and radiologists and other supposed experts insisted that this was absolutely safe. How do they know? They read it in some book or in an AEC publication or they heard it in medical school. In other words, the people who supposedly deal directly with radiation are getting their information from secondary or tertiary sources. In contrast, I am speaking from the basic data on low-level radiation exposures to human beings."

What other basic data exists linking nuclear radiation with disease?

Dr. Thomas Najarian's research into the deaths of 1,722 people who serviced nuclear submarines at the Portsmouth Naval Shipyard in New Hampshire offers still more evidence of low-level radiation dangers.

MORTALITY FROM LEUKÆMIA AND CANCER IN SHIPYARD NUCLEAR WORKERS

THOMAS NAJARIAN
Department of Medicine, Boston University School of Medicine, Boston, Massachusetts, U.S.A.

THEODORE COLTON
Department of Biostatistics, Dartmouth Medical School, Hanover, New Hampshire

Summary A review of death certificates in New Hampshire, Maine, and Massachusetts for 1959–77 yielded a total of 1722 deaths among former workers at the Portsmouth Naval Shipyard where nuclear submarines are repaired and refuelled. Next of kin were contacted for 592. All deaths under age 80 were classified as being in former nuclear or non-nuclear workers depending on information supplied by next of kin. With U.S. age-specific proportional cancer mortality for White males as a standard, the observed/expected ratio of leukæmia deaths was 5·62 (6 observed, 1·1 expected) among the 146 former nuclear workers. For all cancer deaths, this ratio was 1·78. Among non-nuclear workers there was no statistically significant increase in proportional mortality from either leukæmia or from all cancers. The excess proportional leukæmia and cancer mortality among nuclear workers exceeds predictions based on previous data of radiation effects in man.

INTRODUCTION

THIS study was prompted by a case referred to T.N. The patient was a 63-year-old male with pancytopenia and splenomegaly. Bone-marrow biopsy and splenectomy with electronmicroscopy confirmed hairy-cell leukæmia. The patient had been a nuclear welder at the Portsmouth Naval Shipyard (P.N.S.) from 1959 to 1965. The shipyard reported that his total radiation exposure was about 1–2 rem for his 6 years of nuclear work. The patient mentioned that some of his fellow nuclear workers (all younger than he) had died.

Follow-up studies on people exposed to ionising radiation—notably, survivors of the Hiroshima and Nagasaki A bombs, radiologists, Marshallese Islanders, and patients exposed to X-rays for medical purposes—are remarkably consistent in the estimates they yield of the dosage effects of radiation in causing disease. One summary of radiation effects on man[1] estimates that an extra total lifetime dose of 0·1 rem above natural background radiation, if given to the entire U.S. population, would cause about 100 extra cancer deaths per year for about 20 years after the exposure.

Little work has been done on people occupationally exposed to chronic, low levels of radiation and to radioactive materials. High internal radiation doses—after inhalation or ingestion of radioactive materials or absorption of contaminants through cuts in the skin, for example—could cause tissue damage which would be poorly predicted by external gamma ray detectors. Mancuso et al.[2] studied 3520 deaths among former nuclear workers at the Hanford Works in Richland, Washington, and estimated that the radiation dose necessary to double mortality from neoplasms of the reticuloendothelial system and leukæmia was less than 10 rem.

Can the results of studies on A-bomb survivors and persons exposed to medical X-rays be applied to occupational exposure to radioactivity? We have studied proportionate mortality from cancer and leukæmia in a group of workers in the U.S. Naval Nuclear Propulsion Program.

Dr. Najarian concluded that these workers, who had radiation doses not over the supposedly "safe" limit, had six times the rate of cancers of the U.S. population.

The increased numbers of cancer and leukæmia deaths among Naval nuclear shipyard workers seem out of proportion to predictions based on prior knowledge of the effects of ionising radiation in man. Previous data suggest that 50–100 rem doubles leukæmia mortality and 300–400 rem doubles the number of total cancer deaths. Radiation records from the shipyard were not available to us, but radiation doses seem to have been well within national occupational safety standards. Information provided by 50 past and present P.N.S. nuclear workers suggested total radiation doses of less than 10 rem lifetime. Within the Naval Nuclear Propulsion Program the mean radiation exposure for the industrial workers at risk (which includes the shipyard workers) was 0·211 rem annually.[4] The nuclear workers at the P.N.S. had six times the proportional mortality of leukæmia and twice the proportional mortality for all cancers expected for U.S. White males of the same age-groups. These increased figures were found with radiation doses that probably averaged less than 10 rem total lifetime exposure as measured by workers' film badges.*

Dr. Sternglass recently has been studying cancer rates in areas near nuclear plants. The following is part of the paper, "Cancer Mortality Changes Around Nuclear Facilities in Connecticut," which he presented at a Congressional Seminar on Low-Level Radiation on February 10, 1978, in Washington, D.C.

*From *The Lancet,* May 13, 1978.

A detailed study of cancer statistics in Connecticut and nearby New England indicates that cancer mortality increased sharply around two large nuclear reactors in south-eastern Connecticut in direct relation to the measured pattern of accumulated levels of strontium-90 in the local milk. Cancer rates increased most strongly closest to the Millstone Nuclear Power Station located in Waterford where the measured strontium-90 levels reached their highest values, with lesser rises being observed for areas with lower values of strontium-90 in the milk located at increasingly greater distances in every direction away from the Millstone Plant, known to have released the largest amount of radioactive gases ever officially reported for any nuclear plant in the United States.[(1)]

The Haddam Neck plant started to operate in 1968 and the Millstone plant followed in 1970. Between this time and 1975, the most recent year for which detailed data are available, the cancer mortality rate rose 58% in Waterford where the most heavily emitting Millstone plant is located, 44% in New London five miles to the north-east, 27% in New Haven, 30 miles to the west, and 12% for the State of Connecticut as a whole. Rhode Island, whose border is only 20 to 30 miles east of these two plants rose 8%, Massachusetts some 70 miles to the north-east rose 7%,New Hampshire some 120 miles north-east rose only 1%, while for the State of Maine more than 200 miles in the same direction, the cancer death rate actually declined by 6% during the same period.[(2)]

An examination of the radiation doses received by the population drinking the milk in Waterford and nearby New London using the accepted methods recommended by the International Committee on Radiation Protection indicates that the accumulated doses to the bones of children over the period 1970 to 1975 reached values of about 640 millirads from the milk and other food produced in the area, and about 320 millirads to the bone-marrow.[(3)]

This must be compared with a dose of some 2 millirads to the bone-marrow from a typical chest x-ray, so that the very radiation sensitive bone-marrow of children in the New London area received the equivalent of some 160 chest x-rays in the course of 6 years of their most sensitive period of growth and development.

Since bone-marrow type of leukemia is well known from studies of the Hiroshima and Nagasaki A-Bomb survivors to be induced by radiation, and since measurements of the bones of both children and adults have shown a high correlation with levels of strontium-90 in the milk, one is led to conclude the probable existence of a direct causal relation between the abnormally high levels of strontium-90 in the milk near the two Connecticut Nuclear plants and the pattern of cancer changes in Connecticut and nearby New England.

This conclusion is further supported by the fact that the types of cancers that rose most strongly in the Connecticut area are exactly those types that have been found to be most sensitive to radiation in earlier studies as classified by the International Committee on Radiation Protection. Thus, the types of cancers that increased the most in the time available so far were cancers of the respiratory system, which rose 25%, breast cancers, which increased 12%, and cancers of the pancreas, which rose 32%. Since the peak of cancer mortality for respiratory cancers did not occur among the uranium miners until some 7 to 12 years after the onset of irradiation, it is to be expected that further rises in lung cancer will take place in the next five years.[6][7]

Still another observation supports the conclusion that the sharp local rises in cancer in Connecticut are connected with the localized releases of airborne radioactivity from defects in the nuclear fuel, comes from the evidence that the increases were largest for those who were simultaneously exposed to the highest concentrations of other known cancer promoting pollutants, such as industrial chemicals, dust ,pesticides, sulfates, nitrous oxide and other air-pollutants, both in the area where they live and in the working place.

Such synergistic effects are well-known for the case of uranium miners, where the mortality due to lung cancer is some 5 to 10 times greater for miners who only inhaled the radioactive gases but did not smoke while the rate was 50 to 100 times greater for those who did.[6]

Thus, the combined action of airborne radioactivity and ordinary pollution would be greatest for those who live and work in the most polluted viroments, who have the lowest socio-economic status and therefore also the poorest medical care, so that they do not receive the benefit of early diagnosis and treatment. It follows that such synergistic effects involving radioactive and other forms of pollution would be expected to affect most heavily the poorest portion of the population, and this is indeed found to be the case in Connecticut.

Thus, while the total number of cancer deaths increased 15% for the white population of the state as a whole, between 1970 and 1975, this number rose 51% for the non-white or predominantly black population.

Furthermore, in accordance with the greater airborne dust and pollution in chemical factories and other heavy manufacturing, mining and construction activities employing men, the greatest increase in the number of cancer cases during the time the radioactive gases were added to the existing pollutants took place for non-white males, namely by the very large amount of 77%. Thus, the observed pattern of cancer mortality changes in Connecticut and nearby New York and New England since the onset of airborne releases by the two large nuclear plants all fit the expected behavior for radiation - related cancers observed in numerous earlier animal experiments and large-scale epidemiological studies carried out over the last thirty years by many scientists all over the world.

Additional important studies include:

- The work of Dr. Carl Johnson of the Colorado Department of Health showing higher cancer rates among 500,000 people living downwind of the U.S. government's Rocky Flats nuclear installation near Denver. Dr. Johnson found a 140 per cent higher testicular cancer rate in men, sixty per cent higher throat and liver cancer, and forty per cent higher leukemia, lung and colon cancer than the national rate.
- Numerous studies showing up to five times the normal rate of lung cancer among Navajos who mine for uranium. Dr. Joseph Wagoner, director of epidemiological research for the National Institute for Occupational Safety and Health, declares "far too many Navajos have needlessly died" of lung cancer.
- Research by Dr. Samuel Milham of the Washington Public Health Service showing an abnormal rate of cancer fatalities (twenty-five per cent) among workers at the Hanford nuclear facility.
- Research by Dr. Rosalie Bertell of Roswell Memorial, connecting radiation exposure with accelerated rates of aging. She has calculated that one rem exposure is "equivalent to one year of natural aging."

"We greatly underestimated the biological hazard of small amounts of radioactivity in the environment," says Dr. Sternglass, by thinking "there might be a safe threshold dose below which essentially no observable health effects would exist." The underestimation, he says, is "anywhere from 100 to one thousand-fold especially for the developing infant in utero, so that the existing cost-benefit calculations are no longer valid."

But the government's policy does not change. Instead it has adopted a concept of "as low as practicable" radioactive emissions. It admits the medical effects of radioactivity (though still underestimating them significantly) and puts a price-tag of $1,000 per "man-rem"–equivalent either to one person receiving one rem of radiation or 1,000 people each receiving .001 rem (one millirem)–as part of a "cost-benefit" ratio for nuclear power.

Can you explain this cost-benefit ratio?

For each nuclear plant proposed, the AEC and now its successor agency, the Nuclear Regulatory Commission, calculates the "radiological consequences" at $1,000 per man-rem. Here is the NRC staff estimate of the cost to health and longevity of a nuclear power facility the Long Island Lighting Company sought to build at Jamesport:

10.4.2.5 Radiological costs

The Nuclear Regulatory Commission has adopted amendments to Appendix I of 10 CFR Part 50. Appendix I sets forth numerical guides for design objectives and limiting conditions for operation to meet the criterion "as low as practicable" for radioactive material in light-water-cooled nuclear power reactor effluents.

Appendix I may require in some instances that a cost-benefit analysis of additional radwaste systems and equipment that could reduce radiation dose to the population reasonably expected to be within 50 miles of the reactor be performed. In this cost-benefit analysis, the values of $1,000 per total body man-rem and $1.000 per man-thyroid-rem are required to be used.

Using conservative assumptions, the staff has estimated an upper-bound integrated exposure (total body and thyroid) to the U.S. population due to the operation of the Jamesport Station to be 130 man-rems (Sect. 5.4.2.3). Based on $1.000 per man-rem. the annual cost of this exposure is $130,000. This additional cost is negligible when compared to the total annualized cost of operating Jamesport ($272 million at a 70% capacity factor) and when compared to the annualized cost differential between Jamesport and a coal-fired alternative ($183 million). Although the specific assessment of compliance with the maximum organ doses and sequential cost-benefit assessment for specific radwaste augments (if required) will not be completed for some time, the cost of additional equipment, if any is determined to be necessary, would not contribute a significant amount to the overall cost of the facility. Thus, the staff concludes that the radiological cost of the Jamesport Station does not materially affect the overall cost-benefit balance.

How was the $1,000 per man-rem price reached? Dr. Reginald Gotchy, a member of the "radiological impact assessment" section of the Nuclear Regulatory Commission, explained that he consulted with the Federal Aviation Agency, the National Highway Safety Administration of the Department of Transportation and the Social Security Administration to get their actuarial figures for injury, illness and death and factored into a composite of these figures what the NRC sees as the impact of radiation on life. Dr. Gotchy said that the Federal Aviation Agency has been using a figure of $422,000 as the loss from an "aircraft crash fatality" and values an injury at $58,000. He said these figures were derived from settlements after air crashes. He said the National Highway Safety Administration rates an auto crash fatality at $201,000 and "permanent and total disability" from an auto accident at $260,000, a larger amount than for death, he said, because of the cost of extended medical care for disability. Dr. Gotchy said the Social Security Administration rates a cancer death at $40, 000 and a death "from accidents, poison and violence at $164,000."

He could not account for the difference here. He said the Social Security Administration figures a "non-fatal cancer, cancer morbidity" at $38,000.

He said the NRC's compilation of the cost of a death based on these numbers was in a range of $127,000 to $367,000, "about a quarter of a million dollars." But, he said, it was a hard figure to pinpoint. For instance, some of the other projections, Dr. Gotchy said, "have cost put in there for suffering" while some do not. Further, radiation has a genetic impact which the other calculations do not assess. He said the NRC has calculated the financial impact of genetic mutations caused by radiation as $38 per man-rem.

"The $1,000 per man-rem would include everything known to man and a few other things I'm able to figure out," said Dr. Gotchy. But, he continued, "you get somebody like Karen Silkwood who was killed on her way to testify and, God, she got how many million dollars awarded from Kerr-McGee without even having proven her case. So you don't know what the hell's going to happen in a hearing, you just don't know. That's based on a lot of considerations other than the value of what she would have earned in her lifetime or maybe ten others."

Further, said Dr. Gotchy, "some people may think their life is worth more than $250,000. Well, if you ask a person what their life is worth right now, how much would you take for your life, I'm sure you'd get a wide range of answers from those who didn't care whether they lived or died to those who really enjoy life and would part with it very dearly."*

This report, made for the EPA in 1977, tells in even more detail how a price tag is put on the medical consequences of nuclear power:

*Interview with Dr. Gotchy, April 15, 1980.

EPA 520/4-77-003

Considerations of Health Benefit-Cost Analysis for Activities Involving Ionizing Radiation Exposure and Alternatives

A Report of

Advisory Committee
on the
Biological Effects of Ionizing Radiations

Assembly of Life Sciences
National Research Council
National Academy of Sciences

U.S. ENVIRONMENTAL PROTECTION AGENCY
Office of Radiation Programs
Criteria and Standards Division

Scope

Given that decisions are required and that rationality demands analysis before decision, this section will be concerned with an examination of the process of benefit-cost analysis as it relates to ionizing radiation. This radiation affects society in a variety of ways. The effects of the radiation on living organisms are deleterious. It may kill, it may cause cancer, and it may cause genetic damage. An important feature of the radiation at low-dose levels is that the effects are random and are predictable only in a statistical sense. If 1,000,000 adults receive 100 rads of general body radiation, we can predict on the basis of certain assumptions that over a period of 10 years duration of risk about 2,000 of them will develop leukemia (1). We cannot, however, predict who will get leukemia and who will be spared.

Increasing relative scarcity of traditional sources of energy is probable within the coming decades. Nuclear energy has the potential for providing civilization with energy for several centuries but with attendant unwanted radiation exposure. Hence, a document which deals with benefit-cost analysis for activities involving radiation exposure must consider energy sources and power production.

Consideration of Benefit-Cost Analysis for Achieving Lowest Practicable Levels of Ionizing Radiations

Some of the biological effects of radiation will be distributed indefinitely into the future, affecting increasing numbers in future generations. Ordinarily, future costs and benefits of a project are weighted by an exponentially declining discount factor. This is justified on various grounds, such as individual preferences for consumption now, rather than later, and the opportunity cost of capital employed in the project. The consequence in this case is that much of the damage to health entailed by operation in the near future of a system of nuclear power plants, occurring as it will in the relatively distant future, will have its economic cost washed away by discounting. At normal rates of discount of from five to ten percent, not many years are required for this effect to take hold. A human life lost in 1985 because of exposure in 1975 when discounted at 7% is worth, at present, only half as much as a life lost now. Accordingly, the question we face is, can the time stream of benefits, and perhaps more importantly, of costs, be evaluated by means of standard discounting procedures? Or ought we to give special consideration to the very unusual time distributions? The question is even more difficult than that faced in the intra-temporal case, i.e., in attempting to deal with the effects of a project on the distribution of welfare within a single time period or generation. The reason is that many, if not most, of those affected by the decision in the former case, for example on whether to go with the breeder reactor, will not have participated in the decision. How are their interests to be represented, if at all? Is it fair for the present generation to impose the associated radiation load on future generations? We certainly do not have definitive answers to these questions, but they are important, and for this reason, we feel they ought to be raised, at least (7).

If it is known that operation of the project will entail a net change p, $0<p<1$, in the probability of an adverse health effect for each member of the exposed population, then the expected health cost to each can be written as $E(c) = p \sum_{t=0}^{\infty} \frac{Y_t}{(1+r)^t}$, where Y_t represents foregone earnings in period t, and r is the discount rate. The aggregate expected health cost is the sum over all members $E^n(c) = p \sum_{j=1}^{n} \sum_{t=0}^{\infty} \frac{Y_t^j}{(1+r)^t}$, where the number of members is n. Refinements allowing the probability of the adverse effect to vary with age, date of exposure, degree of exposure, and so on, are easily incorporated, as in the measure $\sum_{j=1}^{n} \sum_{t=0}^{\infty} \frac{p_j Y_t^j}{(1+r)^t}$, which introduces different probabilities for each individual.

Another measure of the value of life, or avoidance of injury, that has been suggested (16) is derived from the amount an individual pays to insure against a loss, say his life. For example, if he is willing to pay a $100 premium in a situation in which the probability of loss of his life is .001, it is inferred that he values his life at $100.000 (=$100/.001).

As Mishan (17) has demonstrated, there are problems with each of these measures (and others, essentially refinements of them) apart from their arbitrariness. First, the foregone earnings measures assume that the only thing that matters to an individual or to society is the (reduction in) size of the Gross National Product (GNP). No allowance is made for the loss of utility due to pain, injury, or death. This omission is particularly serious in the case of an elderly or retired person, one for whom all remaining Y_t terms are zero. A somewhat similar problem arises in connection with the life insurance calculation: a person with no dependents might not be willing to pay anything for insurance yet still set a value on his own life.

From preliminary discussion, two tentative conclusions might be drawn. 1) Even if we can legitimately accept (or ignore) the distributional effects of a program involving radiation exposure, including the inter-generational effects, measurement of the expected value of the costs of the exposure will not capture the full value of the costs, due to the risk preferences of the affected individuals. 2) The unusual time distribution of the costs and their potential magnitude raise serious questions about the appropriateness of following standard practice, discounting all benefits and costs and looking only at their present values.

Ethics and Benefit-Cost Analysis

The general thesis in the use of benefit-cost analysis as one of the bases in the making of decisions affecting the public is that such analysis provides one of the objective evaluations on which such decisions should be based, in preference to subjective bases such as those which may derive from intuition, ideology, or political pressures.

In order to compare costs and benefits adequately in benefit-cost analysis, it is necessary to express them in common, comparable terms. Monetization is virtually the only way to arrive at common, comparable terms for summation of various kinds of benefits and summation of various kinds of costs ("costs" including risks to health and life and other usually non-monetized detriments). The value of goods or services is usually defined by "the market" and usually represents the social consensus of the value of the commodities, except perhaps for governmental or private monopolistic price regulations or externalization of costs. Even things which cannot really be valued adequately in the market place, such as a human life or a scenic view, can be assigned a monetary value based upon what people normally pay for them in various ways and circumstances.

Not the least of the arguments in favor of monetization of costs and benefits in benefit-cost analysis is that it is nearly universally accepted in our society. However, there have been severe criticisms of benefit-cost analysis because of the materialistic implications of monetization (18).

Can't we rely on Geiger counters and other measurement devices to warn us about dangerous radiation?

Not necessarily. A very big question arises whether mechanical radiation readings give an accurate reading of the intake of radioactivity by life.

Dr. Sadao Ichikawa has done considerable work in the measurement of radiation through the delicate little flowering plant, *Tradescantia*, or, as commonly known, the spiderwort. Radiation turns the large blue cells of the spiderwort stamen hairs pink. The color change, a result of mutation, can be observed twelve to thirteen days after the plant's exposure to radiation. A clear reading of the extent of radiation can be taken by counting or scoring through a microscope the number of cells that have changed color on the stamen hairs. The spiderwort, stresses Dr. Ichikawa, a geneticist at the University of Kyodo, Japan, displays in days effects of radioactivity which would show in humans only after an incubation period of years.

Mechanical readings of radiation, Dr. Ichikawa emphasizes, measure only "external exposure," not what is breathed in or ingested. Internal exposures can be more significant than one-time external exposures since living organisms "incorporate" and "concentrate" radioisotopes.

Spiderwort plants have responded, under test conditions near nuclear plants, with mutation frequencies fifty times higher than would be expected from the radioactive levels measured by dosimeters. "The present dosimeter method of monitoring environmental radiation can hardly be regarded as efficient from the biological point of view," the geneticist stresses. Spiderworts are now being used as radiation monitors around facilities in the U.S., Japan and Europe. They are being called the "people's radiation monitor." The flower is "the most excellent test system ever known for low-level radiation," says Dr. Ichakawa.

What can be done about radioactive contamination in the body?

The effects of radiation are cumulative and irreversible. Once someone has been exposed to radiation there is nothing that can mitigate its effects. As noted earlier, the U.S. has begun stockpiling potassium iodide pills to be used in the wake of a nuclear plant accident. These would put stable iodine into the body and, the theory is block the radioactive iodine. However, there are many more fission products than radioactive iodine, and potassium iodide has serious side effects.

According to the *Physician's Desk Reference*, potassium iodide carries these risks:

POTASSIUM IODIDE ℞ **1071**

Contraindications: Known sensitivity to iodides is a contraindication to the use of potassium iodide. The drug should not be administered in the presence of acute bronchitis.
Warnings: There have been several reports, published and unpublished, concerning nonspecific small-bowel lesions (consisting of stenosis with or without ulceration) associated with the administration of enteric-coated thiazides with potassium salts. Such lesions may occur when enteric-coated potassium tablets are given alone, with non-enteric-coated thiazides, or with certain other oral diuretics. These small-bowel lesions have caused obstruction, hemorrhage, and perforation. Surgery has frequently been required, and deaths have occurred.

Precautions: Occasionally, persons are markedly sensitive to iodides, and care should be used in administering the drug for the first time. Iodides should be given with great caution, if at all, in the presence of tuberculosis. Because of the possible development of fetal goiter, iodide should be administered with caution to pregnant women.
Adverse Reactions: Thyroid adenoma, goiter, and myxedema are possible side effects. Hypersensitivity to iodides may be manifested by angioneurotic edema, cutaneous and mucosal hemorrhages, and symptoms resembling serum sickness, such as fever, arthralgia, lymph node enlargement, and eosinophilia. Iodism or chronic iodine poisoning may occur during prolonged treatment. The symptoms of iodism include a metallic taste, soreness of the mouth, increased salivation, coryza, sneezing, and swelling of the eyelids. There may be a severe headache, productive cough, pulmonary edema, and swelling and tenderness of the salivary glands. Acneform skin lesions are seen in the seborrheic areas. Severe and sometimes fatal skin eruptions may develop. Gastric disturbance and diarrhea are common. If iodism appears, the drug should be withdrawn and the patient given appropriate supportive therapy.

This EPA guide on nuclear plant accidents says this about potassium iodide:

Manual

of

Protective Action Guides

and

Protective Actions

for

Nuclear Incidents

September 1975

Environmental Protection Agency
Office of Radiation Programs
Environmental Analysis Division
Washington, D.C. 20460

1.6.3.5 Prophylaxis (Thyroid Protection)

The uptake of inhaled or ingested radioiodine by the thyroid gland may be reduced by the ingestion of stable iodine. The oral administration of about 100 milligrams of potassium iodide will result in sufficient accumulation of stable iodine in the thyroid to prevent significant uptake of radioiodine. The main constraint in the use of this means of thyroid protection is that potassium iodide is normally administered only by prescription and would have to be distributed in accordance with State health laws. Potassium iodide as a prophylaxis is only effective if the exposure of concern is from radioiodine and only if the stable iodine is administered before or shortly after the start of intake of radioiodine. All emergency workers for areas possibly involving radioiodine contamination should receive this kind of thyroid protection, especially if appropirate respirators are not available. The cost constraint would not be significant for potassium iodide itself, but the cost for administering this material should be considered, including the cost of testing emergency workers for sensitivity to iodine prior to issue or use.

The use of stable iodine as a protective action for emergency workers has been recommended by EPA, but only in accordance with State health laws and under the direction of State medical officials as indicated above. However, the efficacy of administering stable iodine as a protective action for the general population is still under consideration by government agencies and should not be construed to be the policy of EPA at this time.

And we're back to trying to run away again:

T_M, the time required for people to prepare to leave, depends on such parameters as:

(1) Is the family together?

(2) Rural or urban community? Some farms or industries require more shutdown time than others.

(3) Special evacuations - special planning effort is required to evacuate schools, hospitals, nursing homes, penal institutions, and the like.

(4) There will be some people who will refuse to evacuate.

The best time for T_M for an urban family together might be 0.2 to 0.5 hours, while to shut down a farm or factory might take hours.

The evacuation travel time, T_T, is related to:

(1) Total number of people to be evacuated.

(2) The capacity of a lane of traffic.

(3) The number of lanes of highway available.

(4) Distance of travel.

(5) Roadway obstructions such as uncontrolled merging of traffic or accidents.

The total number of people to be evacuated depends on the population density and affected area. It is an advantage if good planning can keep the area and thus the number of people to as small a value as possible, or possibly to evacuate one area at a time so that the number of people on the move at one time is within the capacity of the roads.

"I am one of those scientists who find it very difficult to see how the human race is to get itself much past the year 2000 unless it makes a drastic change in the way it is going about things," says Nobel Award-winning biologist, Dr. George Wald. "If you were to read in the newspapers tomorrow that astronomers had a shocking piece of information for us, they had just found another star is going to collide with the sun and that would be curtains, we'd have eight months more to go and, finished—why—heavens above! You could put on your best clothes and go dancing in the streets—that's cosmic, that's fate. You could go out with dignity. But the thought of a self-extermination of the human race, bringing along with it much of the rest of life on this planet—for what? . . . It is so trivial, it's so ghastly ignoble as to be, I think, intolerable, altogether unacceptable."

"I call this a lethal society," says Dr. Wald who calls for "the closing down of all nuclear power plants tomorrow."

He warns: "Time is very short and unless we can take our lives into our own hands, unless we can repossess our country, unless we can begin to have our government work for us, disaster lies ahead."

Says Dr. Caldicott: "In view of the threat that nuclear technology poses to the ecosphere, we must acknowledge that Homo sapiens has reached an evolutionary turning point. Thousands of tons of radioactive materials, released by nuclear explosions and reactor spills, are now dispersed through the environment. Nonbiodegradable, and some potent virtually forever, these toxic nuclear materials will continue to accumulate, and eventually their effects on the biosphere and on human beings will be grave: many people will begin to develop and die of cancer; or their reproductive genes will mutate, resulting in an increased incidence of congenitally deformed and diseased offspring—not just in the next generation, but for the rest of time. An all-out nuclear war would kill millions of people and accelerate these biological hazards among the survivors: the earth would be poisoned and laid waste, rendered uninhabitable for aeons."

CHAPTER FIVE

Radioactive Waste

What kind of radioactive waste is left from nuclear power?

Virtually everything involved in nuclear power but the electricity ends up as radioactive waste, even the plant itself. That's why nuclear plants cannot be used after thirty years. They become saturated with radioactivity and too hot with radiation to handle, unlike conventional power plants, some of which have remained in service for over a century.

Further, waste is not really the word to describe the principal materials we are talking of: highly-unstable, radioactive elements which must absolutely be confined and prevented from getting into the biosphere because they cause injury, cancer and death.

This waste is not like ashes from a fire, not just residue but actively poisonous material. The price of breaking up the atom is the creation of lethal twins of safe substances in nature which the human body cannot tell apart. These long-lived man-made poisons must not escape, from the time they are made as fission products in a reactor, through the millions of years some must be stored while they decay and lose their radioactivity. They and what they come in contact with must be forever isolated from life.

Not only are great spans of time involved but gargantuan amounts of poison. Every year one nuclear plant operates, it generates thirty tons of high-level radioactive waste, a speck of which can kill.

Then there is what's defined as the "low-level" radioactive waste produced, although it's only low-level in comparison—so much of it that the U.S. Environmental Protection Agency has calculated that there will be a billion cubic feet of it, if the American nuclear program continues, by the year 2,000. That will be enough, says the EPA, to cover a four-lane highway one foot deep from coast to coast.

Further, many radioactive wastes are gases that can never be retained. They are sent out of the smokestacks of nuclear plants and given off by the mountains of "tailings" left over from the milling of uranium fuel. For each nuclear plant, 4.6 million cubic feet of radioactive mill tailings are produced annually.

Here is how a panel representing many U.S. government agencies recently summarized the radioactive waste situation:

TID-28817 (Draft)
Dist. Category UC-70

Report to the President by the Interagency Review Group on Nuclear Waste Management

October 1978

DRAFT

Washington, D.C.

Waste consists of radioactive species of almost all chemical elements; some contain naturally occurring radioactive materials and others contain man-made radioactive materials; the wastes exist as gases, liquids, and solids. Yet for all their variety, radioactive wastes have one thing in common: as long as they remain highly radioactive, they will be potentially hazardous. This potential hazard results from the fact that exposure to and/or uptake of radioactive material can cause biological damage. In man, it can lead to death directly through intense exposure and a variety of diseases, including cancer, which can be fatal. In addition, radioactive material can be mutagenic, thereby transmitting biological damage into the future.

The central scientific fact about radioactive material is that there is no method of altering the period of time in which a particular species remains radioactive, and thereby potentially toxic and hazardous without changing that species. Only with time will the material decay to a stable (nonradioactive) element. The pertinent decay times vary from hundreds of years for the bulk of the fission products to millions of years for certain of the actinide elements and long-lived fission products. Thus, if present and future generations are to be protected from potential biological damage, a way must be provided either to isolate waste from th biosphere for long periods of time, to remove it entirely from the earth, or to transform it into nonradioactive elements.

While proposals have been advanced through the years about how radioactive wastes might be safeguarded, there has been no solution—and most likely there can never be. How can anything be safeguarded on this earth for millions of years?

Hannes Alfven, a Nobel laureate in physics, has said "there does not seem to be any existing, realistic project on how to deposit radioactive waste." There are the many millenia to be concerned about—and Neanderthal man only appeared on earth 75,000 years ago—and "the deposit must be absolutely reliable as the quantities of poisons are tremendous. It is very difficult to satisfy these requirements for the simple reason that we have had no practical experience with such a long term project. Moreover, permanently guarded storage requires a society with unprecedented stability." Where radioactive waste is put, there can be no "riots or guerilla activity, and no revolution or war. . . . The enormous quantities of extremely dangerous material must not get into the hands of ignorant people or desperados. No acts of God can be permitted."

Radioactive waste is "the main drawback to nuclear development," says the European Nuclear Commission.

And as you read this the radioactive wastes mount up: the spent fuel rods and the mill tailings, the by-products of fuel preparation, the equipment and clothing and tools contaminated in the nuclear fuel cycle and the plants themselves—legacies of poison we leave now for all future generations.

How is radioactive waste being safeguarded now?

It isn't. Some of it, boiling wildly from its own radioactivity and destined to remain fiery for years upon years, is in steel tanks—with useable lives of twenty to thirty years. At the now defunct reprocessing facility in West Valley, New York there is a threat of massive leakage from such tanks.

Much of it is in what is called "spent fuel storage pools" at nuclear plant sites, essentially swimming pools where the water must constantly keep circulating to dissipate the heat of the radioactive waste—which otherwise can erupt with even grimmer consequences than a nuclear plant accident, because more fission products are involved, many times the two tons in a plant. This is what is called a "loss-of-water" spent fuel storage pool accident, and can be set off by a leak in the pool or a breakdown in the pumps which circulate the water. The latter is particularly feared in the event of a severe reactor accident, when personnel would be forced to flee for their lives, leaving the storage pool unattended for months, perhaps years.

A nuclear explosion involving radioactive waste is also possible because of the large amounts of plutonium in the waste. With melting of the waste due to loss-of-water, the large amounts of plutonium could separate out, forming a critical mass. This could happen in a pool or in the trenches in the ground into which radioactive waste has also been dumped.

Indeed, a nuclear explosion in a trench at a radioactive waste site in the Soviet Union in 1958 is regarded as the nuclear accident with the largest loss of life so far. Such a "criticality incident" was also feared at the American nuclear waste storage facility in Hanford, Washington.

Further, around the world tens of thousands of 55-gallon barrels full of nuclear waste have been dumped into the oceans. Many of these barrels have now cracked open and, in the Pacific just off San Francisco, three-to-four foot mutant sponges have been growing from the cracks. The seabed sediment in the area is heavy with plutonium. Here is an Environmental Protection Agency photo of one of the mutant sponges:

In the beginning of the history of nuclear power, the United States offered to be the radioactive dumping ground for much of the world—as it tried to sell the world on its reactors and nuclear technology. The

U.S. agreed to take back spent fuel and this currently is the subject of a major lawsuit in America.*

Here are some examples of so-called radioactive waste disposal:

At the U.S. Government's nuclear facility in Hanford, more than 450,000 gallons of high-level liquid radioactive waste has leaked over the past 30 years into the soil and into the water table underlying the facility, which lies along the Columbia River. Within only a two month period in 1973, undetected by technicians and supervisors, some 115,000 gallons of waste, including plutonium strontium-90 and cesium-137, leaked from tanks which the U.S. Geological Survey had described a decade before as a "potential hazard," their "structural life not entirely known."

And, at Hanford, a 570 square-mile site where seventy-five per cent of U.S. nuclear waste is stored, the practice for years was to dump radioactive waste—including liquid plutonium—into trenches with concrete sides and a top *but no bottom.*

Here, from a government report entitled "Radioactive Wastes" is the theory:

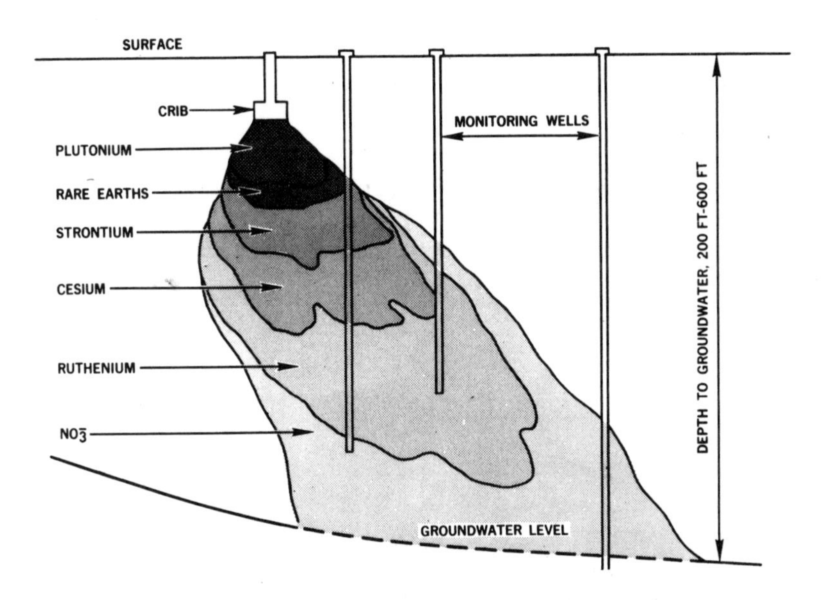

*Virginia Sunshine Alliance and Truth in Power, Inc. et al, vs. U.S. Nuclear Regulatory Commission, Federal District Court, Washington, D.C.

Obviously, that radioactive material was not going to stay put. And, in 1973, the U.S. Atomic Energy Commission declared that in one trench–Z-9–"due to the quantity of plutonium contained in the soil . . . it is possible to conceive of conditions which could result in a nuclear chain reaction."

Meanwhile, the Federal Water Pollution Control Agency has described the Columbia as "the most radioactive river in the world." And the Sierra Club has charged that Hanford's radioactive wastes "pose a serious threat to the food chains of the whole Pacific."

In 1979, after Stephen Stalos, an environmental physicist at Hanford, charged there was an official policy to cover up news about leaks in tanks at the facility, the Department of Energy made this investigation.

U.S. Department of Energy

Office of Inspector General

Report on

Alleged Cover-Ups of Leaks of Radioactive Materials at Hanford

January 22, 1980

IGV-79-22-2-231

The Energy Department's Office of Inspector General declared:

> We decided that our IG inquiry would be most useful if it focused on that portion of Mr. Stalos' expressions of concern relating to alleged cover-ups of leaks of radioactive nuclear waste at Hanford that Mr. Stalos believed were being perpetrated by the nuclear waste management contractor, Rockwell International, and the contract administration organization, the Department's Richland Operations Office. In broad, Mr. Stalos charged that the organizations concerned were following a policy of not announcing tank leaks although a Hanford policy that was promulgated in 1973 requires all radiation leaks at the site to be promptly and publicly announced to local and regional news media. Mr. Stalos said that the considerations causing him to believe that there was such a policy of cover-up included a series of statements made to him by Richland and DOE employees.

But, said the Office of Inspector General, it couldn't even get the information at Hanford.

> A Disquieting Episode Relating to the Reclassification Study
>
> The draft report of the Rockwell tank farm surveillance group that was distributed in March of last year recommended that six Questionable Integrity tanks be reclassified as Confirmed Leakers. During the previous four years, no tanks had been so classified. The draft report would therefore have been of obvious interest to our office, and of obvious relevance to our inquiry.
>
> Yet when a representative of our office working on this investigation visited Hanford during the period May 29 through June 16 of last year, Rockwell middle management officials with whom he talked

who knew about this report did not mention its existence to our colleague. We learned of it through a person whose identity we do not plan to reveal since we wish to protect the individual from possible reprisals. We requested a copy of it and one was furnished to us. As we were given to understand, a meeting involving Rockwell middle management officials had earlier been held and during that meeting the question of whether the existence of this draft report should be volunteered to our office was discussed. Also, according to what we were told, it was decided that the report's existence should not be volunteered to us, but that we should be provided with a copy of the report if we found out about it and asked for it.

Concluded the Department of Energy:

The word "cover-up" evokes pictures of people devising strategies and tactics aimed at concealing things which ought not to be concealed. But in the case of Hanford, had there been any officials desiring to minimize publicity about tank leaks, they would have had no real need to engage in conduct which might be considered questionable. This is because Hanford's existing waste management policies and practices have themselves sufficed to keep publicity about possible tank leaks to a minimum.

To the Office of Inspector General of the DOE, the Hanford facility is not only covering up news to the public about tank leaks but this major nuclear waste facility is even keeping information from the government which pays the bill.

In the American West there are now 130 million tons of radioactive mill tailings in huge piles, blowing away in the wind, contaminating the water. For many years this waste, made up of U-238, was distributed to construction companies for use in concrete with which homes, sidewalks, schools, factories and other projects were built.

Left underground the uranium does not endanger life because it is blanketed by the earth for its 4.5 billion year half-life. Uncovered, it emits radioactivity into the atmosphere—particularly radon gas—for an equal length of time. Dr. Chauncey Kepford, a chemist, has argued (without being disputed) at federal nuclear hearings including the licensing hearings for Three Mile Island, that the radon emitted from the mill tailings left from the ore to fuel one nuclear power plant for just one year will ultimately cause a million deaths. In Grand Junction, Colorado, where many structures were built with mill tailings, it was discovered that people were receiving the equivalent of 500 chest X-rays or ten rems of radiation each year from the radiation emanating from the floors of their homes.

It was the nuclear industry and U.S. government's theory that radioactive waste could be put into salt mines, "stable geological formations" and dry, it was maintained.

In 1971, the U.S. Atomic Energy Commission began excavating a salt deposit in Lyons, Kansas for the purpose.

Here is the plan:

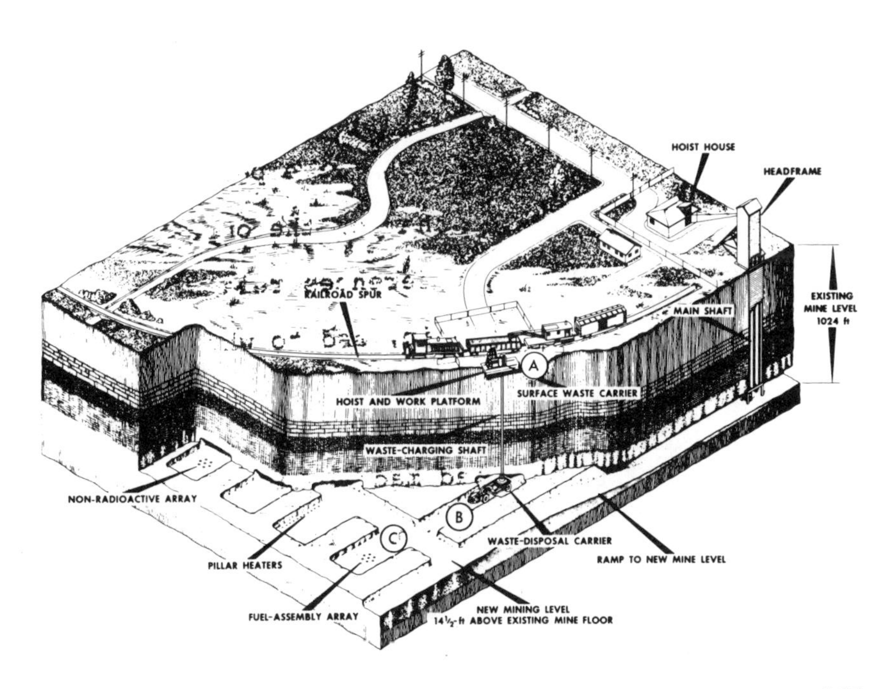

What wasn't considered was the heat of high-level radioactive wastes—over 1,000 degrees Fahrenheit. The hot canisters would plummet through the salt. And salt, it became clear, was not all that dry. Declared the U. S. Environmental Protection Agency, even the driest salt deposits reveal "significant amounts of water in fluid inclusions." Radioactive waste canisters, it said, are "likely to be bathed in water soon after emplacement." Salt is highly corrosive, EPA stressed. "It is likely that the canister could be breached within scales of a decade or less."

"The mystique has built up that salt is dry and it's OK," says David Stewart of the U.S. Geological Survey. "Salt is not dry and it's not OK." The government found out that the mine it had dug in Kansas, at a cost of millions, was adjacent to an active salt mine, where water was long used for hydraulic mining, and that holes from the old mine were penetrating the new mine. In April 1980 the U.S. government gave up on it latest attempt to place nuclear waste in salt, abandoning its "Waste Isolation Pilot Project" which was to use salt caverns outside Carlsbad, New Mexico. Some $90 million was spent.

"Either the federal bureaucracy is to a large extent incompetent," the Natural Resources Defense Council has declared, "or the radioactive waste disposal problem is considerably more difficult than has been publicly admitted by the nuclear power industry. To a large degree both explanations are supported by a careful examination of the record."

It has been thought that radioactive waste might be encapsulated in glass. But, it turned out, glass quickly melts at the temperature of radioactive wastes. "Glass is soluble and it's leachable, it's what you would do if you wanted to maximize activity in the geological environment" says Dr. William Luth, a Stanford University geochemist.

As the California Resources and Development Commission put it in a 1978 report, there is just "no safe method" of disposing of radioactive wastes and it is questionable one can be developed in the future. Therefore:

> . . . the *evidence* indicates that it is not prudent to continue siting nuclear powerplants based on an optimistic assumption that waste management technologies to handle nuclear waste will be developed and scientifically demonstrated . . . there are substantial scientific gaps which *preclude proceeding on the basis of faith* that all the attendant risks and issues will be resolved.

This chart from a U.S. Nuclear Regulatory Commission report, "Alternative Processes for Managing Existing Commercial High-Level Radioactive Wastes," tells the story beyond the "basis of faith." For every plan other than holding wastes in steel tanks, "Technology has not been demonstrated."

TABLE 2.2. Advantages, Disadvantages and Status for NFS Waste Management Alternatives Based on ERDA Processes

Alternative	Advantages	Disadvantages	Current Technological Status
Liquid Storage	No offsite shipping Fair retrievability Amenable to change in form if necessary	Potential for leakage Tank life short compared to necessary storage time. Higher probability for dispersion during natural acts, e.g., flood and earthquake, than for monolithic solid. Continuous surveillance, active control, and emergency response capabilities are required for centuries.	Fully operational for storing neutralized high-level radioactive waste.
Conversion to Cement	Low-leaching solid product Good retrievability Easily transported	Difficult to change to alternative form if necessary Technology has not been demonstrated. Need to store large volume of residual salt Potential for container to pressurize Salt product would contain significant quantities of water (up to 20%).	Process is under active laboratory development and conceptual design of facilities for application to ERDA neutralized high-level radioactive waste.
Shale Fracturing	Low-leaching solid product No offsite shipping Immediate placement in geologic formation Process has been field tested using intermediate-level radioactive waste. Not susceptible to natural acts on earth's surface, e.g., glaciation, denudation, flood, tornados.	Site verification required Criteria for disposal of long-lived alpha emitters not established Waste is not retrievable. Not possible to change to alternative form Requires pipeline transport of liquid high-level waste for distance of about 1 mile Technology has not been demonstrated for NFS waste.	Full-scale demonstration for ERDA neutralized low- and intermediate-level radioactive wastes at Holifield National Laboratory.
Shale Cement	Low-leaching solid product Key portions of process use state-of-the-art technology Good retrievability Easily transported	Large volume of waste-grout mixture must be stored. Potential for container to pressurize Difficult to change to alternative form if necessary Technology has not been demonstrated.	Process is in the conceptual stage.
Calcination	Good retrievability Easily transported Product would be thermally and radiolytically stable	Technology has not been demonstrated for NFS waste. Difficult to change to alternative form if necessary Without glassification, calcine product is quite leachable.	Fully operational using acidic, low-sodium ERDA high-level radioactive waste.
Aqueous Silicate	Reasonably low-leaching solid product Good retrievability when stored in canisters Easily transported	Product would contain significant quantities of water (up to 35 wt%). Canisters must resist internal corrosion associated with high water content. Difficult to change to alternative form if necessary Large volume of aqueous silicate product must be stored. Potential for canister to pressurize Technology has not been demonstrated.	Aqueous silicate product has been produced on a laboratory scale using ERDA neutralized high-level radioactive wastes.
Conversion to Glass	Low-leaching solid product Glass product would be thermally and radiolytically stable. Good retrievability Easily transported	Technology has not been demonstrated. Difficult to change to alternative form if necessary	Glass product has been produced on a laboratory scale for ERDA neutralized radioactive high-level waste. Has been demonstrated on pilot plant scale for acidic high-level waste.
Salt Cake	Amenable to change in form if necessary Fair retrievability	High leach rate Limited thermal and radiolytic stability Product would contain significant quantities of water. Potential for container to pressurize Containers must resist internal corrosion. Technology has not been demonstrated for NFS waste. Sludge is not treated.	Fully operational units now processing ERDA neutralized high-level radioactive wastes.

The theory of dumping waste in a "stable geological formation"—that will be stable for the millions of years necessary—is a greatly questionable notion in the first instance. "Our knowledge of geology," stresses the Sierra Club, "is not sufficient to guarantee the stability of any area with the certainty demanded by the risks involved. We cannot predict what physical changes a specific location will undergo over time with the accuracy needed . . . . We know major climatic changes can occur more than once in 500,000 years. The last Ice Age, for example, was only 10,000 years ago. We have no way of predicting such events." And the organization adds that "it is irresponsible to accelerate production of long-lived wastes before we know what we are going to do with them."

"Earth scientists can indicate which sites have been relatively stable in the geological past, but they cannot guarantee future stability," declared Dr. Newell Trask, the leader of a team of USGS geologists and hydrologists which investigated radioactive waste disposal in 1978. "We still don't know enough about such major geological events as earthquakes and climate changes to predict their occurrence and effect for the next thousand or hundreds of thousands of years."

"What we are doing," says Lorna Salzman of Friends of the Earth, "is storing waste without knowing whether it will in fact be isolated, and none of us will be around to know if the scientists are right or wrong."

Even the general manager of Electricite de France, Marcel Boiteux, admits that "it is natural to have scruples about leaving behind such a present for future generations."*

As Illinois Attorney General William Scott has said:

> . . . on the whole problem of the storage of hazardous wastes (the state government thinks) of it so far in terms of . . . some vague idea of thorium dust, spent fuel rods, instead of thinking of people with birth defects, brain damage, people dying horrible deaths of cancer.
>
> . . . If the people who were exposed to . . . radioactivity would all of a sudden drop over right away, then you could relate to it. But the fact that the . . . radioactivity will cause bladder cancer 15 or 25 years from now, does not immediately surface, so the problem is not that dramatic.
>
> . . . We have a time bomb ticking already. **

**Nuclear Power*, edited by John Lambert, agenor, Brussels, Belgium, 1975.

**Testimony before a subcommittee of the House Committee on Government Operations, September 22, 1977.

Can radioactive wastes be rocketed into space?

Former U.S. Energy Secretary James Schlesinger suggested this but the costs have been projected at billions of dollars a shot and it is deeply feared that if the rocket explodes the entire global atmosphere would be condemned to perpetual radioactive fallout. Other speculative and highly-criticized schemes have been to put radioactive wastes in the ice caps and in cracks in the ocean floor.

What happened at the radioactive waste dump in the Soviet Union in 1956?

It was an explosion of nuclear wastes stored underground, contaminating an area the size of Rhode Island, the kind of accident that was feared could happen at Hanford.

Nuclear Disaster in the Urals by Soviet scientist Zhores Medvedev is a book (published by W. W. Norton & Company, 1979) about the disaster, what Dr. Medvedev describes as "the biggest nuclear tragedy in peacetime that the world has known" and producing "the largest radioactively contaminated ecological zone in the world."

The U.S. Central Intelligence Agency, at the request of Ralph Nader's Critical Mass Energy Project in 1977, released documents describing the aftermath of the accident. In one, "a strange uninhabited and unfarmed area" was described. Reported a traveler in the region: "Highway signs along the way warned drivers not to stop . . . because of radiation. The land was empty, there were no villages, no towns, no people, no cultivated land, only the chimneys of destroyed houses remained."

Here is a segment of one CIA document:

-3-

APPROVED FOR RELEASE
Date 14 January 1977

Kyshtym

In spring 1958, [redacted] he heard from several people that large areas north of Chelyabinsk were contaminated by radioactive waste from a nuclear plant operating at an unknown site near Kyshtym, a town 70 kilometers northwest of Chelyabinsk on the Chelyabinsk-Sverdlovsk railroad line. It was general knowledge that the Chelyabinsk area had an abnormally high number of cancer cases. To go swimming in the numerous lakes and rivers in the vicinity was considered a health hazard by some people. Food brought by the peasants to the Chelyabinsk market (rynok) was checked by the municipal health authorities in a small house at the market entrance where the peasants also paid their sales tax. How radioactive food was destroyed was unknown to source. Food delivered to the plants, schools, etc., by the kolkhozy and sovkhozy was probably examined by the latter themselves. Until 1958 passengers were checked at the Kyshtym railway station, and nobody could enter the town without a special permit By what authority the permit was issued and why the checking was discontinued in 1958, source was unable to say. In addition, some villages in the Kyshtym area had been contaminated and burned down, and the inhabitants moved into new ones built by the government. They were allowed to take with them only the clothes in which they were dressed.

Hundreds of people were reported killed, thousands contaminated. Hunting and fishing were prohibited in the south and central Urals. Entire lakes and stream systems were saturated with radioactivity. Strontium-90 levels in one lake exceeded the Soviet drinking water standard by a factor of 5,000. One CIA document said people in the region "grew hysterical with fear, with the incidence of unknown 'mysterious diseases breaking out.'" One source wrote of a "terrific explosion" accompanying a massive release of radiation over the area. "Very quickly all the leaves curled up and fell off the trees."

Another report describes a scene in a local hospital. "Some of them were bandaged and some were not. We could see the skin on their faces, hands and other exposed parts of the body to be sloughing off. These victims of the blast were brought into this hospital during the night. It was a horrible sight."

The U.S. government, in this study of the Soviet nuclear accident, declares a "significant area" was involved.

Due to the high population density in this region (the industrial Urals, 95) and the reported level of ^{90}Sr contamination alone (100,101), the incident probably resulted in the evacuation and/or resettlement of the human population from a significant area.

ORNL-5613

Analysis of the 1957-58 Soviet Nuclear Accident

J. R. Trabalka
L. D. Eyman
S. I. Auerbach

ENVIRONMENTAL SCIENCES DIVISION
Publication No. 1445

This is based on the government's discovery that thirty Russian communities are now literally off the map.

Comparisons of high-resolution (1:250,000) maps of the area between Cheliabinsk and Sverdlovsk based on materials produced before (1936-1954) and after the accident (1973-1974), respectively, indicated the deletion of over 30 names of small communities (< 2000 population) within the dashed area of Fig. 1. None of the names of towns and villages shown on the earlier editions within the 70-km-long southwest-northeast running arm of the dashed area appear on the later editions. A somewhat wider zone (10-15 km vs 7 km) runs in a southeasterly direction toward the Sverdlovsk-Cheliabinsk highway, generally along the Techa River; however, names of a few communities still remain in this area. A number of the communities whose names no longer appear had evidently grown to $\geq$ 2000 population size by the late 1950's as their presence on low-resolution atlases (9,106) testifies (Boyevka, Yugo-Koneva, and Russkaya Karabolka in the northeast arm and Metlino and Asanova in the southeast arm of the dashed area in Fig. 1). Further, population centers in other parts of the region appear to have developed extensively in the same period; nowhere else in the Sverdlovsk-Cheliabinsk area has such extensive deletion of community names occurred. Collectively, this information could be construed to indicate the relocation of the human inhabitants from the area in a time frame consistent with the contamination incident.

It seems rather apparent that the Soviet nuclear program has had to contend with a catastrophe involving reprocessed nuclear wastes.

What about the notion of "air-cooled" concrete canisters of waste?

This is yet another scheme. Here's a rendition of it by the Atomic Industrial Forum, Inc., the nuclear industry trade group:

Dr. Richard Webb calculates that because of the huge volume of radioactive waste involved "a land area of the size of Massachusetts" would be needed for the projected twenty-five to seventy-five million canisters, and the amount of concrete needed to make them would be five to fifteen times the amount of concrete used to pave the present system of superhighways in the United States." And, he notes, each canister could undergo a meltdown of the hot nuclear waste inside if the air was blocked.

What about the claim that there is more radioactive waste from the military nuclear program than from nuclear power plants?

In volume, existing commercial wastes equal and will soon exceed military wastes. As a recent Congressional report declared:

> Contrary to widespread belief, the accumulated inventory of fission products generated by the still small U.S. civilian nuclear power industry may already be comparable to that generated in the past by U.S. military nuclear programs.

Most important, commercial wastes are far more toxic than the military wastes; they contain many times more fission products, and at this point are building up much faster than military nuclear garbage.

So how is the government dealing with the thirty tons of high level waste a nuclear power plant produces yearly?

With storage pools for spent fuel at nuclear plants filled to capacity (and the nation's reactors facing shutdown with no more room on-site to store spent fuel) the federal government is now planning to expand three nuclear sites, at West Valley, N.Y., Barnwell, S.C. and Morris, Ill.—and turn them into vast "storage areas" for spent fuel. The spent nuclear plant fuel will be stored in mazes of pools of circulating water. These facilities are to be called "Away-From-Reactor" (AFR) sites. By the middle 1990's, said the Carter administration in early 1980, it hopes that a plan might be devised on what to do with the spent fuel after that.

Would reprocessing help?

What is called "reprocessing"—the extraction of much of the plutonium and unused uranium from nuclear waste for re-use as fuel, by stewing the waste in acid—still leaves everything else, all the other fission products, the bulk of the waste. And huge amounts of radioactive gases would be released during reprocessing–enough krypton-85 that alone could exhaust the allowable exposure of the general population to radioactivity. Further, reprocessing is another case of reality not matching technological fantasy. A few examples: Getty Oil opened its Nuclear Fuel Services reprocessing facility in West Valley, New York in 1966. There was constant pollution of the air, land and water in the messy process and widespread contamination of workers. Many young people were hired briefly as "sponges" or "jumpers"–working on tasks requiring them to absorb maximum radiation doses, after which they were fired.* Getty Oil shut down the facility in 1972 and decided to abandon it in 1976–leaving 600,000 gallons of high-level radioactive waste bubbling in tanks ready to give out and at least $480 million needed, perhaps $1 billion, to deal with the mess. The State of New York was forced to take over the facility. General Electric tried to open a reprocessing facility in Illinois in the early 1970's but gave up after spending $64 million and finding it didn't work. And now a subsidiary of Gulf Oil Company has built and plans to open a reprocessing plant near Barnwell, South Carolina.

What about the nuclear plants themselves? What will happen to them after their thirty-year lifetime?

The nuclear industry and government talk of reactor decommissioning but it is yet another untried, toweringly expensive theory–currently estimated at twenty-five to one hundred per cent of the original cost of constructing a nuclear power plant.

Here is the Comptroller General's report on the matter:

*Some 1,400 men, most of them just over 18, were used as "sponges" at West Valley between 1967 and 1972 when the facility closed. "It was unskilled labor. Sometimes it only involved turning a bolt," said public relations man Steve Sass of Nuclear Fuel Services, in 1979. Dr. Bross said, "One guy would go in, turn a screw a quarter of a turn, then rush out. It was the most callous use of human beings since the slave trade."

REPORT TO THE CONGRESS

BY THE COMPTROLLER GENERAL OF THE UNITED STATES

Cleaning Up The Remains Of Nuclear Facilities-- A Multibillion Dollar Problem

Energy Research and Development Administration
Nuclear Regulatory Commission

The problem of protecting the public from the hazards of radiation lingering at nuclear facilities which are no longer operating needs Federal attention if a strategy for finding a solution is to be developed.

The solution doubtless will be expensive--but the expense should be known so the responsible parties can plan for the inevitable cost. A strategy to clean up these privately and federally owned nuclear facilities, which continue to accumulate, cannot be developed until basic questions on the magnitude of the problem, such as costs, radioactivity, and timing, have been answered.

EMD-77-46

JUNE 16, 1977

CHAPTER 2

FACILITIES AND ACTIVITIES

THAT MUST BE CLEANED UP

While a nuclear activity is ongoing, the materials, equipment, and facilities that come into contact with a nuclear reaction or radioactive material could become contaminated or radioactive. Once the activity is ended, disposing of these items presents special problems. Facilities once used for nuclear activities cannot be abandoned if radioactive materials remain that present a radiation hazard. Structural materials or equipment cannot be recycled if they have been made unsafe by contact with a nuclear activity. A nuclear operations building cannot be reused for other purposes unless radioactive materials and contamination have been removed or reduced to acceptable levels.

Many types of nuclear facilities must be prevented from endangering public health and safety. Each type will have to be handled, or decommissioned, in a different way. A major factor in determining the best way is the nature of the radiation hazard at the facility.

Two types of hazards could be involved in a nuclear facility: induced radioactivity and surface contamination. Induced radioactivity results from a nuclear reaction and is embedded in the equipment or material coming into contact with the nuclear reaction. This induced activity cannot be cleaned-up and can remain dangerous for thousands of years. For this reason, a structure containing induced radioactivity should be dismantled at some point in time. This should be done before the structure begins to deteriorate, thus permitting the radioactivity to enter the environment.

Surface contamination results from facilities or equipment coming into contact with radioactive material. As opposed to induced activity, material having surface contamination can often be cleaned up by scrubbing and washing.

In describing the cleaning-up process, the words decontamination and decommissioning are often used. In this report, decontamination denotes the process of cleaning-up surface contamination. Decommissioning is a term used by NRC and ERDA to indicate the closing or shutting down of a facility with some actions taken to prevent--at least temporarily--health and safety problems. It does not necessarily denote a permanent solution to cleaning-up the facility.

The report declared:

CHAPTER 4

MAJOR QUESTIONS REMAIN UNANSWERED

To begin to grapple with the far-reaching problems of decommissioning requires answers to some basic questions. In our review, we found that the questions listed below and discussed in the following section have not been answered.

--How much will decommissioning cost and who should pay?

--How should nuclear reactors be decommissioned?

--What is the extent of the decommissioning problem for accelerators?

--Are standards needed for induced radiation?

--What should be the limits on acceptable radiation levels?

--What more should States do to plan for decommissioning?

DECOMMISSIONING--HOW MUCH WILL IT COST AND WHO SHOULD PAY?

Privately owned facilities

The total cost to decommission privately owned nuclear facilities in the United States is unknown. Very few studies have been made on the subject. In fact, to the best of our knowledge, only one major study on the cost to decommission commercial nuclear reactors has been done to date, and another NRC-sponsored study is in process.

A 1979 NRC report places hope in "entombment."

3.4 ENTOMBMENT

The operations required for entombment include, basically, chemical decontamination where necessary, and storage, in the containment building below the operating floor level, of as much as possible of the contaminated equipment and material located elsewhere in the power plant. A continuous slab of concrete is then poured above the operating floor level in the containment building, and all wall penetrations below the floor level are sealed. *

*From: "Technology, Safety and Costs of Decommissioning A Reference Pressurized Water Reactor Power Station," U.S. Nuclear Regulatory Commission, 1979.

Essentially, this means blighting the world's landscape with cemented-over billion dollar devices that could be used but thirty years and then must be guarded perpetually . . . or else.

CHAPTER SIX

Economics and Jobs

What about the claim that nuclear power is an inexpensive souce of energy?

This was for a brief period, a hope of the nuclear industry. In 1954 former U.S. Atomic Energy Commission chairman Lewis Strauss spoke of electricity from nuclear power as "too cheap to meter." In fact, it is the most expensive energy source—and its price continues to skyrocket.

This is because of:

CAPITAL COSTS. Nuclear plants turned out to be exhorbitantly expensive to build. Extremely complex anyway, their inherent danger has required all sorts of back-up and emergency systems not required for safe fuel. And as accidents have piled up, more and more systems have been needed to try to deal with the host of problems and dangers. They now cost over fifty per cent more to build than conventional power plants.

CAPACITY. Nuclear plants have been able to operate only a little more than half the time. Their complexity, the pressures under which they operate, the deterioration caused by radioactive bombardment—all of this leads them to be plagued with troubles. And as they get older they are more prone to breakdown, like any machine (and accidents are expected to increase).

FUEL. The price of uranium has quadrupled in the past seven years. As supplies shrink, it will become an even more expensive fuel source. Indeed, a severe worldwide uranium shortage is projected soon.

OPERATION. The costs of trying to keep nuclear plants going has turned out to be enormous, not only because of their complexity and chronic breakdowns but also due to the difficulty of repairs. Workers must be found who will labor in areas hot with radioactivity. They must work quickly for a very short time (often just a few minutes) for as they work they are undergoing maximum radiation absorption or "burn out." On one repair of piping at Consolidated Edison's Indian Point I plant, which could have been made in short order by a few repair people in a conventional power plant, it took eight months and 700 workers; it cost $1 million.

And then there are the many "hidden costs," particularly those involving the billions upon billions of dollars in governmental subsidies nuclear power has received—all paid through your taxes.

These extra costs include:

INSURANCE SUBSIDY. Without the Price-Anderson Act limiting the liability of a utility for a nuclear plant catastrophe to a fraction of the expected cost, utilities would have to pay $23.5 million per plant per year for insurance on the private market, former Pennsylvania Insurance Commissioner Herbert Dennenberg has calculated. But no insurance company would even take the risk.

ENRICHMENT. The government plants which were set up to "enrich" or concentrate the U-235 in uranium to make highly fissionable material for use in atomic bombs are now simultaneously and primarily used to prepare uranium for nuclear power plants; the government has charged only a third to a half of what was figured out to be the "commercial rate" for doing this. Some thirty-five per cent of the enriched uranium produced at these plants is shipped to other nations, also at a cut-rate to fuel their American reactors. These billion-dollar uranium enrichment facilities, incidentally, consume three per cent of America's electricity, substantially diminishing any contribution to the electric supply by nuclear power.

TAXPAYER SUPPORTED RESEARCH AND DEVELOPMENT. Many billions have been spent on government sponsored research into civilian nuclear power at the string of national laboratories and associated facilities (Oak Ridge National Laboratory, Brookhaven National Laboratory, Savannah River National Laboratory, Argonne National Laboratory, Puerto Rico Nuclear Center, Idaho National Engineering Laboratory, among others) set up across America by the U.S. Atomic Energy Commission for that purpose. These are research and development facilities for the nuclear industry, compliments of the American taxpayer. The major share of U.S. research and development in energy, now under the U.S. Department of Energy, continues to go into nuclear power.

An extensive analysis of costs to consumers was made in 1978 by the Critical Mass Energy Project and the Environmental Action Foundation.* They surveyed America's 100 largest electric companies

*"Nuclear Power and Utility Rate Increases," Washington, D.C., June 30, 1978.

and concluded that "the use of nuclear power to generate electricity has usually resulted in higher utility rates for consumers. These findings have grave implications for consumers in many parts of the country where nuclear reactors are now under construction or are in the planning stages . . . . The data suggests that long-held claims of consumer savings is largely a myth. Rather, this study shows . . . a strong correlation between the use of nuclear power and the rise in electric rates."

Meanwhile, the reactor manufacturers are having a rough time—even with the lavish government hand-outs. As the 1978 Congressional report,* "Nuclear Power Costs," notes:

> Dr. Bertram Wolfe, general manager of nuclear engineering program operations of General Electric Co. in San Jose, Calif., said that, without a sustained average of 16 to 20 orders a year, "the present structure of the industry is not going to be able to hold out." [199]
>
> Westinghouse has said its commercial nuclear division stopped growing about 2 years ago. (It received only four domestic orders for nuclear plants in 1975 and none since then.) [200]
>
> General Electric reportedly faces losses of $500 million, Babcock & Wilcox faces a $200 million loss, and Atomic General, a Gulf Oil subsidiary, was forced to withdraw from the nuclear business in November, 1975. Nor are foreign nuclear vendors faring any better, according to the same report. The British nuclear industry has received no orders for plants since 1970 and West Germany and Japan are facing virtual nuclear moratoriums because of increasing public opposition. The West German company Kraftwerk Union did not make a profit from the time it was set up in 1969 until 1976. The recent profits were due to two big orders from Iran and Brazil. Of the 10 domestic orders it received since 1973, seven are not being built because of public opposition. The Canadian nuclear power agency, Atomic Energy of Canada Ltd. was expecting losses of $200 million.[201]
>
> If nuclear power is as economically competitive with other energy sources as its supporters claim, why are nuclear vendors facing such immense financial losses?

Other "hidden costs" include governmental support for storage of waste, the uranium depletion allowance—twenty-two per cent compared to ten per cent for coal—and these, listed in the Congressional "Nuclear Power Costs":

*The Congressional Committee which made the report was chaired by Representative Leo Ryan, killed in Jonestown, Guyana in 1979.

Other costs associated with nuclear power which have a significant bearing on the cost of that power will also be examined. These include:

"Phantom taxes," or tax costs anticipated by utilities and charged to customers, but never required to be paid by those utilities to the U.S. Treasury. These "taxes" netted utilities $1.5 billion in 1975,[8] and $2.1 billion in 1977.[9]

The "Fuel Adjustment Clause," which allows a utility to adjust its rates automatically to account for changing fuel expenditures. Rate-setting hearings are bypassed either permanently or temporarily. The clause has also served to insulate utilities from the risks associated with nuclear plant operations by allowing utilities to adjust their rates to reflect almost any unexpected occurrences, such as plant breakdowns. The cost of substitute power from coal- or oil-fired plants is charged to customers through this clause, instead of being absorbed by utility stockholders.[10]

"Construction Work In Progress" (CWIP) allows utilites to charge customers now for current construction costs of powerplants. Customers pay more, and over a longer period of time, for electricity that will not be produced until some time in the future when the plants are completed. CWIP boosts electric rates by as much as 15 percent and enables a utility to reduce its need to borrow money by requiring customers to provide capital for new construction. This is especially important for the construction of nuclear plants with their high capital cost.[11]

Then there are the potentially enormous costs associated with the "back end" of the fuel cycle. The costs of virtually indefinite radioactive waste storage and decommissioning of the nuclear plant remain essentially unknown, and, in most cases, have not been factored into the price the present-day consumer pays for nuclear-generated electricity.

"The conclusion that must be reached," declares investment counselor Saunders Miller, a specialist in energy costs, "is that from an economic standpoint alone, to rely upon nuclear fission as the primary source of our stationary energy supplies will constitute economic lunacy on a scale unparalleled in recorded history, and may lead to the economic Waterloo of the United States."

"Never in history has there been a plan to have the entire economy of a large industrial nation so dependent upon a technology built on so fragile an economic foundation," he says. He calculates the planned American nuclear energy program costing $5.8 *trillion* in plants and support facilities. "This compares with the estimated $160 billion figure for Vietnam, which merely culminated in double-digit inflation . . . . On a national scale, the deleterious effects upon the economy would make the perturbations and inflation caused by the Vietnam War pale into insignificance."*

Barron's, the financial publication, in a review of Miller's 1978 book, *The Economics of Nuclear and Coal Power*, concluded: "If the top executives of every electric utility in the nation–not to mention the utility regulators–read Saunders Miller's new book . . . most nuclear plant construction projects would be scrapped."

These trillions of dollars would be going to an energy form–electricity–which can now and in the future only be a small fraction of total energy. Electricity can never be a principle energy source.

As energy economist Vince Taylor explained in his 1979 report for the U.S. Arms Control and Disarmament Agency entitled *Energy: The Easy Path:*

The Limited Potential of Nuclear Power

Nuclear electric power will be able to make, at best, only a marginal contribution to future energy supplies in the next twenty-five or so years. This marginal contribution will not significantly lessen dependence on the Middle East nor reduce anticipated growth in consumption of oil and gas and, thus, will do little to avert the energy shortages that, in the conventional view, threaten to occur near the end of the century. No complex analysis is required to understand the limitations of nuclear-electric power. Electricity is a special, very expensive form of energy, and economic considerations tend to restrict the use of electricity to those applications where its special properties justify its premium price. As a result, in the major industrial nations, the share of electricity in total end-use energy ranges from 10 percent in the United States to 15 percent in Japan.

Because electricity is so ubiquitous and noticeable, for instance in lights, TVs, refrigerators, air conditioning, hi-fis, its small share in total energy consumption probably comes as a surprise to many people, who intuitively believe that electricity constitutes a major part of the total. The electrical

**The Economics of Nuclear and Coal Power*, Praeger Publishers, New York, 1976.

share is so small because electricity is used almost entirely for lighting and for driving stationary motors. Only a very minor portion of electricity is used for heating and transportation, reflecting the high cost of electricity relative to oil and gas for these uses. Yet, heating and transportation are the dominant users of energy in all industrial economies.

If nuclear power is to reduce the share of energy provided by scarce fossil fuels beyond the point possible by replacing oil and gas in present electrical generation, electricity must in the future perform functions that are now performed by directly consumed fossil fuels. A detailed survey of the possibilities for such substitutions shows them to be severely limited.

- Ten percent of directly-consumed fuels were used as chemical feedstocks for highway paving, offering no possibilities for nuclear electricity.

- Thirty percent of directly-consumed fuels were used to provide process heat in industry. Switching to nuclear electricity would cost over 4 times as much as continued use of fuel oil.

- Motor transport accounted for about 35 percent of directly-consumed fossil fuels. With present technology, electric cars are generally inferior to gasoline automobiles in cost, performance, durability, and range. Considering that they must compete with new, far more efficient generations of gasoline-powered autos that are in prospect, electric cars seem unlikely to take over more than a minor share of the transportation market during this century.

- The remaining, major use of fossil fuels, space heating of buildings, represented about 20 percent of the total. Although heat pumps can reduce the economic disadvantage of nuclear electricity for heating buildings, the present cost of this method of heating is still about twice as much as oil per unit of delivered heat. Further, if heat pumps were to provide one-fourth of the estimated heating requirements of new dwellings constructed between 1975 and 2000, their contribution in 2000 would amount to only about one percent of 1975 end-use energy consumption.

In sum, there is little prospect for a substantial increase in the present, small share of energy consumed in the form of electricity. A similar look at the structure of energy use in Japan and the major countries of Europe yields a similar conclusion.

In the 1960's the two principal manufacturers of nuclear power plants, Westinghouse and General Electric, all but gave away reactors in what were called "turnkey" deals: at a fixed price G.E. and Westinghouse would build plants for utilities and take the losses (which totalled more than $850 million for the two firms). At the same time the U.S. government began a "Cooperative Power Reactor Program" distributing $260 million to utilities to build nuclear plants. The utilities were thus lured into a nuclear "bargain."

Their greatest concern from the outset was their liability to pay for catastrophic damage which, it was clear from the start, nuclear plants are capable of inflicting.

But with the passage of the Price-Anderson Act (1957) which relieved them of most of the liability, the utilities figured that the nuclear business might be profitable for them. That is because of the way utilities are set up financially.

The U.S. government allows them to be monopolies. In the absence of competition their profits are set by regulatory commissions on the basis of capital investments. It's a "cost-plus" game for utilities: the more they buy and build, the more money they can make.*

Do they offer savings to consumers? No.

For it's the ratepayers who finally pay the bill for the utilities' power plants. Utilities using nuclear power to generate electricity normally charge the highest rates.

Richard Morgan of Environmental Action, author of *Nuclear Power: The Bargain We Can't Afford,* told the House subcommittee which assembled "Nuclear Power Costs":

*See Ron Lanoue, *Nuclear Power Plants: The More They Build, The More You Pay*, 1976.

"What I suggest is a dose of free enterprise. Get rid of all the financial subsidies and let the utilities take their own risks on nuclear plants. Then I think we will see nuclear power die a quiet death."

Meanwhile, there is the impending uranium shortage.

As the Atomic Industrial Forum admitted in 1977:

> The light water reactors coming on line through the remainder of this century will likely consume all estimated economic U.S. uranium supplies, proven and potential, during their lifetimes. *

"The best scientific evidence available indicates when the sum of all U.S. uranium reserves are developed, no more than the equivalent of 100 large nuclear power plants" in addition to those already built in the U.S. "could ever be fueled," declares John Berger, former energy projects director of Friends of the Earth.

He adds: "Moreover, the developing worldwide uranium shortage indicates that the tremendous uranium requirements of the U.S. nuclear industry are unlikely to be met by importing uranium from abroad. Reactor-grade uranium is an extremely scarce world resource and world uranium supplies are not at our disposal; a great many other countries have laid claim to them. Furthermore, international uranium producers are organizing into an effective marketing organization or cartel, and dependency on a bloc of uranium producers does not bring the U.S. any closer to energy independence than does dependence on foreign oil producers."

The Congressional panel's report declares:

> Having been subject to the whims of the OPEC countries in the past, the United States should not leave itself vulnerable to another cartel that could have power over energy supplies.

"The economics of nuclear power are bad and getting worse," says energy consultant Charles Komanoff. "In my judgement, no utility executive with an accurate perception of the costs of nuclear power

*From "Assessment of the Nuclear Fuel Cycle," Atomic Industrial Forum Committee on Fuel Cycle Policy, 1977.

and a sincere desire to minimize costs would propose ordering a new nuclear plant."

The author of three books on power costs, counselor on energy to seven state governments, to the federal government and to many local governments, Komanoff calculates that nuclear power now costs an average of twenty-two per cent more than electricity from coal, throughout the U.S.

Broken down by U.S. region by kilowatt cost, the figures are striking. Note that in the Northwest, nuclear power is forty-nine per cent in excess of coal costs.

Region	Coal Cost	Nuclear Cost	Excess Nuclear Cost
Northeast	5.5¢	5.6¢	1%
East North Central	4.6¢	5.4¢	18%
South Atlantic	4.6¢	5.2¢	11%
South Central	4.2¢	5.2¢	23%
West North Central	4.3¢	5.4¢	25%
Mountain & Pacific NW	3.6¢	5.4¢	49%
California	4.5¢	5.6%	25%

An even cheaper alternative than coal is solar power.
"Nuclear Power Costs" recognizes this.

If the Federal Government spent only a small portion of what it has already spent on nuclear power development for the commercialization of solar power, solar generated electricity would be economically competitive within five years, in the view of many experts.

Indeed, John O'Leary, a deputy secretary of the U.S. Department of Energy, testified before the House panel in 1977:

> A solar power breakthrough will solve the energy crisis once and for all . . . A viable plan to use the inexhaustible solar power source is reachable within 5 years.

"Nuclear Power Costs" stresses the versatility of solar power:

> Solar energy is not only energy derived directly from the sun, it also encompasses wind power, water power and biomass. The largest portion of current commercial solar usage is of biomass—that is, bioconversion or the production of fuels from wood, dung, crop residues or other agricultural material which store energy from the sun. This accounts for 90 percent of all energy used in many Third World countries.
>
> The world's oceans are another source of solar power. Ocean thermal electric conversion (OTEC) plants could use the ocean as a free solar collector and storage system and are unaffected by whether the sun shines or not. Because the ocean's temperature hardly varies, these plants could be a steady, round-the-clock source of power.
>
> About one-fifth of all energy used throughout the world now comes from solar resources. By the year 2000, these renewable sources could provide 40 percent of all needed energy and 75 percent by 2025.

As the Club of Rome, a group of European thinkers and statesmen, says: "We can choose the solar alternative to achieve a permanent, clean source of abundant energy in a fully ordered and economically feasible global transition. Solar energy will prove to be economically the least expensive and socially the most affordable path."

Another healthy option is energy efficiency.
Notes "Nuclear Power Costs":

> More than half the current energy produced in the United States is wasted. For the next 25 years the United States could meet all its new energy needs simply by improving efficiency. The energy saved could relieve the immediate pressure to commit enormous resources to energy sources such as nuclear power, before all alternatives have been fully explored. Reducing energy demand through conservation would be safer, more reliable and less polluting than producing energy from other sources. Most importantly, a strong energy conservation program would save consumers billions of dollars a year.[327]
>
> Conservation means many things. It means conserving by such methods as more efficient use of energy through insulating buildings, refitting furnaces, and cutting electric demand during peak hours. But it also means recovery—recovery of energy from garbage and waste and reusing industrial steam, by a process known as cogeneration.
>
> The potential energy supplies from these methods are enormous.

For example, the California Energy Commission found that the potential for industrial cogeneration in that State alone could be as much as 6,700 megawatts by 1985.[328] That is equivalent to the generating capacity of about five nuclear plants.

Americans waste more fuel than is used by two-thirds of the world's population. Nowhere is this more evident than in over-air-conditioned, over-heated and overlit American buildings. The World Trade Center in New York City—where one person working at night at his desk must turn on a quarter-acre of lighting—is a perfect example. Those twin buildings use more electricity than the entire city of Schenectady, N.Y.[329]

The American Institute of Architects (AIA) has reported that a commitment to developing energy-efficient buildings could alone save more energy by 1990 than nuclear power is projected to supply at that time even at historical growth rates.[330]

Richard Stein, Chairman of the New York Board for Architecture, in his book Architecture and Energy, shows that energy consumption could be reduced by one-half in buildings operated by the State of New York with a capital expenditure that could be repaid by fuel cost savings in 2 years. Much of the reduction could be achieved through improved insulation, temperature setbacks related to use, and reduced and improved lighting. It is less expensive to improve energy efficiency of existing buildings, he concluded, than to build a powerplant of equivalent capacity.[331]

The manufacture of building materials also offers great savings opportunities. The United States currently uses an average of 1.2 million Btu's to decompose enough limestone to produce a barrel of cement. In European plants, where waste heat from cement kilns is recaptured to preheat the limestone, only 550,000 Btu's are needed per barrel. In addition, Stein calculates that the electricity used in manufacturing unnecessary cement alone amounts to some 20 billion kilowatt-hours a year—roughly equivalent to the electricity consumed by 3 million families.[332]

In view of this, it would appear the United States could reduce its energy consumption by 40 percent or more, without adverse effects on industrial output or individual lifestyles and with the positive effects of increasing employment and reducing inflation and pollution.[333] Furthermore, since prosperous and highly industrialized countries such as Sweden and West Germany consume about 40 to 50 percent less energy per capita than we do, it would appear there is a significant potential for energy savings.

There is no one-to-one relationship between energy use and well-being as energy producers would have us believe. Those who say, "The more energy we use, the better off we are. If we want to be even better off in the future we will have to use even more energy. Energy conservation would mean a poorer America," do not have convincing evidence.

In recent times, as demand for electricity declines, interest rates are high and modifications are required because of the Three Mile Island accident, gloom has pervaded the economic side of the nuclear industry.

"Nuclear Energy: Dark Outlet," was a 1979 report of The Washington Analysis Corporation, a subsidiary of Bache Halsey Stuart Shields. The financial prospects for nuclear power are summarized in the report's opening statement: "Washington Analysis Corporation estimates the overall prospects for the domestic nuclear industry as highly unfavorable."

General Public Utilities, the owner of the Three Mile Island plant, has been brought to the brink of bankruptcy by the accident. A year after the accident, the firm's stock, which had been selling for $17 a share is down to less than $5. It still could not use either of its reactors on Three Mile Island. And to replace the electricity the complex had provided, it has had to borrow heavily to purchase substitute power, creating a severe drain in cash and raising the specter of bankruptcy. In a 1979 annual report, the company's auditors, Coopers and Lylbrand, questioned whether the utility would be able to "continue as a going concern." The company, which has three subsidiaries, Jersey Central Power & Light, Metropolitan Edison and Pennsylvania Electric, has been having difficulty raising its rates as ratepayers have balked at bailing it out.

Three Mile Island was an "eye-opener" to the utilities "who jumped into the nuclear industry in the 1960's . . . . They thought the Price-Anderson legislation was going to save them," says former General Electric nuclear engineer Gregory Minor. But they "never anticipated" the huge costs for buying substitute energy and clean-up in a somewhat less than catastrophic accident.

What about jobs?

Here again, the hype has been one thing, the facts another. As the Congressional "Nuclear Power Costs" concludes:

> Nuclear plants are capital-intensive and thus produce few jobs. Renewable energy sources such as solar and conservation are not capital intensive, and are expected to produce many jobs—500,000 construction jobs for solar hot water installation alone—or three times as many jobs as produced by the nuclear industry.

A detailed study on the issue was completed in 1979 by the Council on Economic Priorities. For two and a half years Long Island, N.Y. was the focus of an analysis (partially funded by the U.S. Department of Energy) comparing "the employment and economic impacts" of nuclear power in relation to solar power and energy efficiency.

A local utility had proposed to build a two-unit nuclear facility. The Council compared the number of jobs that would be generated if the same amount of money went instead to a variety of solar and energy efficiency measures, the technology for which "is readily available" and which would meet the same energy needs. Its study concluded that the solar/energy efficiency path would produce 2.2 times as many jobs locally, for each dollar spent, as the nuclear project.

Only residential solar and energy-saving elements were considered (among them: passive solar systems, solar hot water heaters, weather-stripping, improved wall and attic insulation, night set-back heating, automatic flue dampers, attic fans, insulation of storm windows and doors, hot water tank re-insulation, heat exchangers and the use of energy efficient appliances). "If the same measures were extended to the commercial and industrial sectors of Long Island's economy," said the Council's 300 page report, *Jobs and Energy*, "energy savings and job generation would be far greater." Its analysis shows that "on a dollar-and-cents basis, nuclear power does not make sense."

The Council* stressed that the solar/energy efficiency path created not only more jobs but a better distribution of employment than nuclear power.

Other studies by the states of New York and Massachusetts also have concluded that alternative energy forms produce more jobs than nuclear power.

In their 1977 report, "Jobs and Energy," Environmentalists For Full Employment concurred:

> **A recent perceptive ERDA report has recognized that among other aspects of the nation's energy dilemma, unwarranted "fear of unemployment is a key political fact."[130] This will be true as long as energy monopolies insist on threatening economic depression and unemployment if their expansion of vast, complex, costly and centralized energy systems is not permitted to continue.**
>
> **To be sure, jobs will trickle down as a result of investment in wasteful and dangerous energy systems. ERDA has estimated that the current total employment in nuclear fission**

*Established in 1969 "to disseminate unbiased and detailed information on the practices of U.S. corporations in areas that vitally affect society" and to thus "assure corporate social responsibility."

electric activities is about 80,000 people: mostly engineers, mathematicians, physical and earth scientists, technicians, welders, plus "all other employees."[131] Getty Oil Company's Nuclear Fuel Service Facility, between 1966-1971, employed an average of 1400 temporary workers each year at radioactive "hot spots." To locate, repair and insulate six 4½ inch hot water pipes in radioactive areas of a nuclear reactor, Consolidated Edison Company brought in 1500 welders, each of whom worked 15 minutes until he had received his maximum permissible dose of radiation. Professor Irwin C. Bupp of MIT has calculated that proposed floating nuclear power plants will create jobs beyond those created by land-based nuclear plants—jobs in hull scraping and ferrying.

Promoters of ever-expanding deployment of nuclear and other huge energy systems as the primary means of providing employment and prosperity (such as *Americans for Energy Independence*) try hard to portray those seeking energy efficiency and the commercialization of solar technologies as being against "economic growth . . . workers, the poor, and the disadvantaged."

But it is the energy expansion scenario, which wastes both capital and resources, provides only limited jobs and unreliable energy sources, which causes disease and environmental destruction, which is not in the interests of "workers, the poor and the disadvantaged," or in the interests of anyone else except the large energy corporations themselves. There are more jobs by far—and safer jobs, and there will be greater prosperity more evenly distributed if the nation cuts back significantly on energy waste and moves vigorously toward solar energy. There will also be much less social and political havoc arising from this path to energy sufficiency.

Clearly, those who seek this solution do not seek an era of freezing and starving in the dark. They envision just the opposite: a time of decreasing dependence on foreign countries and on vulnerable and speculative energy systems; a time of abundant jobs and healthier people who live amidst cleaner air and water; and a time when people have greater control over their own lives and more resources with which to obtain the goods and services which make living easier and more enjoyable for all. A fair and equitable transition to a conservation and solar economy, during which no group or class of people will be made to bear the burdens of changing social values and technological innovations, would mean that the entire society would benefit greatly.

Energy corporation supporters try to suggest that the average American has no business getting involved in energy problems and solutions. *Americans For Energy Independence*, for example, advises citizens to defer to "representatives" in government, labor and industry.

Many Americans are ahead of their leaders in understanding the causes of the nation's energy and unemployment problems. They are willing to seek solutions which may not necessarily coincide with corporate myths. They realize that energy efficiency and solar technologies are the methods by which the public can be assured that enough safe energy will be available: that the people will be able to control its production and use; and that there will be sufficient numbers of jobs available in diverse activities throughout a prosperous nation.

In a 1980 report, "Energy And Employment," the group declared that "There is little indication that government policy makers" are considering jobs "in their deliberations. Fundamentally, employment is not considered a major factor in selecting energy policies. Neither are alternative technologies with their excellent job-creating possibilities . . . . Without active constituency pressures it is unlikely that employment issues will be integrated into energy policy making, or that new technologies—with their energy-providing and job-creating potential—will be sufficiently implemented."

Despite the "desperation media blitz . . . that nuclear power was necessary for jobs and economic security," says Ralph Nader, "citizens recognize that it is extremely expensive and employment-inhibiting."

Says Nader: "Even without a major accident or legislated moratorium, atomic power is doomed because economics—including its costs to ratepayers, taxpayers, and future generations—will stop it. The industry may continue to stumble and bumble along, but the electrical demand to justify the industry's own growth projections will not materialize. Alternatives—conservation at first, and eventually solar power—will develop to displace and dissolve nuclear power. One danger is that the electric atom's proponents, suffering from a Vietnam mentality, will attempt to prop up the technology with a massive infusion of more overt government subsidies and incentives. If that happens, the political response will be analogous to Vietnam. The more proponents attempt to prop up the industry, the more visible will atomic power's economic defects and hidden subsidies become, the more political opposition will arise, and the more difficult will it become to pass the next round of subsidies to keep the industry going. In the end, the industry will have finally collapsed and the nation will have recognized that the useless infusion of funds was avoidable."*

*Testimony before a subcommittee of the House Committee on Government Operations, September 19, 1977.

CHAPTER SEVEN

How We Got So Far

How *did* we get so far?

The structure of what became the nuclear industrial establishment—an assemblage of nuclear scientists, government bureaucrats and giant corporations—was born innocently in the 1939 letter from Albert Einstein to President Franklin D. Roosevelt. As earlier noted, Einstein and other scientists, particularly refugees from the Nazis, were fearful that fission, just performed for the first time in Germany, might be turned into a war weapon by the Nazis. The first page of the letter proposed that the U.S. make use of fission to build "extremely powerful bombs of a new type."

The second page of the letter (which is on display at the Roosevelt Library and Museum in Hyde Park, N.Y.) laid out ways to do this:

Albert Einstein
Old Grove Rd.
Nassau Point
Peconic, Long Island

August 2nd, 1939

F.D. Roosevelt,
President of the United States,
White House
Washington, D.C.

Sir:

Some recent work by E.Fermi and L. Szilard, which has been communicated to me in manuscript, leads me to expect that the element uran-

The United States has only very poor ores of uranium in moderate quantities. There is some good ore in Canada and the former Czechoslovakia, while the most important source of uranium is Belgian Congo.

In view of this situation you may think it desirable to have some permanent contact maintained between the Administration and the group of physicists working on chain reactions in America. One possible way of achieving this might be for you to entrust with this task a person who has your confidence and who could perhaps serve in an inofficial capacity. His task might comprise the following:

a) to approach Government Departments, keep them informed of the further development, and put forward recommendations for Government action, giving particular attention to the problem of securing a supply of uranium ore for the United States;

b) to speed up the experimental work, which is at present being carried on within the limits of the budgets of University laboratories, by providing funds, if such funds be required, through his contacts with private persons who are willing to make contributions for this cause, and perhaps also by obtaining the co-operation of industrial laboratories which have the necessary equipment.

I understand that Germany has actually stopped the sale of uranium from the Czechoslovakian mines which she has taken over. That she should have taken such early action might perhaps be understood on the ground that the son of the German Under-Secretary of State, von Weizsäcker, is attached to the Kaiser-Wilhelm-Institut in Berlin where some of the American work on uranium is now being repeated.

Yours very truly,

A. Einstein

(Albert Einstein)

Einstein would later regret what he did. "If I had known that the Germans would not succeed in constructing the atom bomb, I never would have moved a finger." He stressed that "we helped in creating this new weapon in order to prevent the enemies of mankind from achieving it ahead of us, which, given the mentality of the Nazis, would have meant inconceivable destruction and the enslavement of the rest of the world." But, he went on, in 1945, "physicists find themselves in a position not unlike that of Alfred Nobel," the inventor of TNT, who "in order to atone for this, in order to relieve his human conscience, he instituted his awards for the promotion of peace and for achievement of peace. Today, the physicists who participated in forging the most formidable and dangerous weapon of all times are harassed by an equal feeling of responsibility, not to say guilt."

"Since I do not foresee that atomic energy is to be a great boon for a long time, I have to say that for the present time it is a menace," Einstein declared.*

But the 1939 letter and its proposal of how the political configuration to make atomic bombs could be put together—between "the Administration and the group of physicists working on chain reactions in America . . . a person who has your confidence . . . Government Departments . . . University Laboratories . . . co-operation of industrial laboratories" ended up creating a grouping of those who'd have a vested interest in what was supposed to be a wartime crash program.

The "Manhattan Project" was formed, the top-secret World War II atomic bomb production project in America. By 1945, 600,000 people had become part of a two billion dollar program at nuclear facilities that had been quickly built across America, with most of its work—from management to engineering—done by large corporations and major universities and most of its money channeled through them. Making four atomic bombs had become a major part of the U.S economy.

There was Union Carbide at Oak Ridge; Stone & Webster in Chicago; Du Pont at Hanford, Washington; the University of California at Los Alamos; and General Electric and Westinghouse making equipment for the undertaking. It was the "basic wartime policy of General Leslie Groves," (head of the Manhattan Project), "and the Manhattan Project that contracting with a few of the nation's largest and best qualified companies and universities was the most expeditious and

**Out of My Later Years,* Albert Einstein, Philosophical Library, N.Y.

effective way to develop, design and produce atomic bombs," explains a subsequent study by the Brookings Institution of Washington entitled *Contracting For Atoms*.

By the end of the war, many of the people and corporations involved in the wartime program didn't want to see it over and their contracts ended.

Japan surrendered after a uranium-fueled atomic bomb was dropped on Hiroshima on August 6, 1945, killing 100,000 and injuring 100,000 many of whom died subsequently, and after a plutonium-fueled atomic bomb was dropped on Nagasaki on August 9, killing 39,000 and injuring 25,000 many of whom died later of the after-effects. People are still dying in sizeable numbers from the radioactivity unleashed on those two mornings.

At the laboratory where atomic bombs were put together there were now "new . . . pressures," James Kunetka relates in his 1978 book, *City of Fire,** about the Los Alamos Scientific Laboratory and the Manhattan Project. There were now the problems of "job placement, work continuity . . . more free time than work." There was "hardly enough to keep everyone busy, and certainly not the atmosphere of urgency . . . . Without a crash program underway the Laboratory found itself for the first time discouraging overtime, and staff members and their families were encouraged to take accumulated leaves." The word was that "Los Alamos would survive in one form or another" but "much of the spirit promoted by the war and the threat of a Nazi bomb was gone . . . . No one doubted that the government would continue to support bomb research, and that in all likelihood the research would be controlled by a new government agency or commission."

Create a government office for anything, give a corporation a contract for anything—even a wartime exigency—and a vested interest is created. And the Manhattan Project created an extraordinary far-ranging complex of vested interests—a technological empire of precedent-setting proportion.

But how could this technological complex be perpetuated?

Atomic bombs aren't things that easily lend themselves to commercial spin-off.

In the first nuclear reactors lay the clue. They had been built at Hanford to turn uranium into plutonium for bombs—and as a by-product gave off heat. The theory: modify these devices to use their heat to boil water to make steam to turn a turbine to make electricity.

*Prentice-Hall, Inc., Englewood Cliffs, N.J., 1978.

It's a hell of a way to boil water. As Amory Lovins has put it, like "cutting butter with a chain saw." But it was a way to keep the machinery going, to let the army of people and the giant corporations involved in the Manhattan Project continue.

It's as if a way to set off a monsoon was devised during the war and afterwards those with a vested interest in monsoon-making would try to peddle it: monsoons for peace. They'd wash a city clean, the would-be monsoon-manufacturers would say.

Out of the atomic bomb thus came "atoms for peace."

But the notion had to be peddled cleverly to win public acceptance. The enormous risks had to be hidden.

During the war, as a matter of wartime censorship, the Manhattan Project had gained a great deal of experience at concealing the truth and manipulating the public.

An example: the first test of an atomic bomb was widely seen and heard, yet no one (except those in the Manhattan Project) was to know what really had happened. It was July 1945, a month before the atomic bombs would be dropped on Japan. The Alamagordo Bombing range in New Mexico was chosen for the site of the test, code-named Trinity. The Manhattan Project men were unsure what would happen. As *City of Fire* relates, "safety was a second concern. What if radioactive dust drifted over nearby towns? Plans were made for Major T.O. Palmer of the U.S. Army to be stationed north of the test area with 160 enlisted men on horses and in jeeps. Palmer was instructed to evacuate ranches and towns at the last moment if necessary. Another twenty men in Military Intelligence were disguised as civilians and stationed in nearby towns and cities up to 100 miles away. Most of these men were armed with recording barographs to get permanent records of blast and earth shocks. The nearest towns were the most obvious candidates for disaster: San Marcial, San Antonio, Soccorro, Carrirozo, Oscuro, Three Rivers, Tularoso, Alamagordo." As part of the plan, an intelligence officer had been "stationed" in the Associated Press office in Albuquerque, New Mexico "to prevent alarming stories from going out."

The bomb was set off. The Los Alamos Laboratory director, Robert Oppenheimer, viewing the fireball rising, the desert bathed in eerie, blinding white light, the ominous mushroom cloud billowing was struck, he recalled, by the words of the sacred Hindu book, the *Bhagavad-Gita.*

I am become death.
The shatterer of worlds.

For great distances the blast was felt. *City of Fire:*

"The explosion had been seen elsewhere. The first flash of light was seen in Albuquerque, Santa Fe, Silver City and El Paso. Windows had been broken in nearby buildings and had been rattled in Silver City and Gallup. A rancher sleeping near Alamagordo was awakened suddenly with what seemed like a plane crashing in his yard. Wives from Los Alamos on Sawyers Hill saw a great flash of light that lit up the trees and produced a long, low rumble." And "the Associated Press office in Albuquerque soon had a number of queries and reports on a strange explosion in southern New Mexico."

The "precautions" General Grove had taken were put into operation. The intelligence officer stationed at the A.P. office, identified as Phil Belcher, gave the Associated Press the "news release" the Manhattan Project had prepared for the occasion. And the Associated Press obediently ran the following story:

> **Alamagordo, July 16 – The Commanding Officer of the Alamagordo Army Air Base made the following statement today: "Several inquiries have been received concerning a heavy explosion which occurred on the Alamagordo Base reservation this morning.**
>
> **"A remotely located ammunition magazine containing a considerable amount of high explosives and pyrotechnics exploded.**
>
> **"There was no loss of life or injury to anyone, and the property damage outside of the explosives magazine itself was negligible.**
>
> **"Weather conditions affecting the content of gas shells exploded by the blast make it desirable for the Army to evacuate temporarily a few civilians from their homes."**

"New Mexico newspapers ran the story in different versions, and the story appeared in a number of radio shows," *City of Fire* notes. "No further word was issued by the Alamagordo Base."

The first atomic bomb was detonated in a blast stirring cities and towns through America's southwest, and there was no difficulty in "managing" the news about it.

This continues in the story of nuclear power to the present day. Managing information flow, intimidating and quashing press inquiry, not letting the citizenry know what's going on through the heavy use of public relations techniques would mushroom like the cloud from that first blast. Indeed, considering the lethal effects involved, its threat to the survival of life, a broad cover up has been central to the nuclear undertaking, in order that it might continue.

The political vehicle permitting the Manhattan Project's work to go on and expand was the Atomic Energy Act of 1946.

Here it is:

[PUBLIC LAW 585—79TH CONGRESS]
[CHAPTER 724—2D SESSION]
[S. 1717]

AN ACT

For the development and control of atomic energy.

Be it enacted by the Senate and House of Representatives of the United States of America in Congress assembled,

DECLARATION OF POLICY

SECTION 1. (a) FINDINGS AND DECLARATION.—Research and experimentation in the field of nuclear chain reaction have attained the stage at which the release of atomic energy on a large scale is practical. The significance of the atomic bomb for military purposes is evident. The effect of the use of atomic energy for civilian purposes upon the social, economic, and political structures of today cannot now be determined. It is a field in which unknown factors are involved. Therefore, any legislation will necessarily be subject to revision from time to time. It is reasonable to anticipate, however, that tapping this new source of energy will cause profound changes in our present way of life. Accordingly, it is hereby declared to be the policy of the people of the United States that, subject at all times to the paramount objective of assuring the common defense and security, the development and utilization of atomic energy shall, so far as practicable, be directed toward improving the public welfare, increasing the standard of living, strengthening free competition in private enterprise, and promoting world peace.

ORGANIZATION

SEC. 2. (a) ATOMIC ENERGY COMMISSION.—

(1) There is hereby established an Atomic Energy Commission (herein called the Commission), which shall be composed of five members. Three members shall constitute a quorum of the Commission. The President shall designate one member as Chairman of the Commission.

(2) Members of the Commission shall be appointed by the President, by and with the advice and consent of the Senate. In submitting any nomination to the Senate, the President shall set forth the experience and the qualifications of the nominee. The term of office of each member of the Commission taking office prior to the expiration of two years after the date of enactment of this Act shall expire upon the expiration of such two years.

RESEARCH

SEC. 3. (a) RESEARCH ASSISTANCE.—The Commission is directed to exercise its powers in such manner as to insure the continued conduct of research and development activities in the fields specified below by private or public institutions or persons and to assist in the acquisition of an ever-expanding fund of theoretical and practical knowledge in such fields. To this end the Commission is authorized and directed to make arrangements (including contracts, agreements, and loans) for the conduct of research and development activities relating to—

(1) nuclear processes;

(2) the theory and production of atomic energy, including processes, materials, and devices related to such production;

(3) utilization of fissionable and radioactive materials for medical, biological, health, or military purposes;

(4) utilization of fissionable and radioactive materials and processes entailed in the production of such materials for all other purposes, including industrial uses; and

(5) the protection of health during research and production activities.

PRODUCTION OF FISSIONABLE MATERIAL

SEC. 4. (a) DEFINITION.—As used in this Act, the term "produce", when used in relation to fissionable material, means to manufacture, produce, or refine fissionable material, as distinguished from source materials as defined in section 5 (b) (1), or to separate fissionable material from other substances in which such material may be contained or to produce new fissionable material.

(b) PROHIBITION.—It shall be unlawful for any person to own any facilities for the production of fissionable material or for any person to produce fissionable material, except to the extent authorized by subsection (c).

(c) OWNERSHIP AND OPERATION OF PRODUCTION FACILITIES.—

(1) OWNERSHIP OF PRODUCTION FACILITIES.—The Commission, as agent of and on behalf of the United States, shall be the exclusive owner of all facilities for the production of fissionable material other than facilities which (A) are useful in the conduct of research and development activities in the fields specified in section 3, and (B) do not, in the opinion of the Commission, have a potential production rate adequate to enable the operator of such facilities to produce within a reasonable period of time a sufficient quantity of fissionable material to produce an atomic bomb or any other atomic weapon.

(2) OPERATION OF THE COMMISSION'S PRODUCTION FACILITIES.—The Commission is authorized and directed to produce or to provide for the production of fissionable material in its own facilities. To the extent deemed necessary, the Commission is authorized to make, or to continue in effect, contracts with persons obligating them to produce fissionable material in facilities owned by the Commission.

"Profound changes in our present way of life" are projected by the act as happening–and they would not be a matter of choice for the people. The entity set up by the law, the Atomic Energy Commission, would have special powers to organize those changes: to both promote nuclear development and somehow simultaneously regulate it. Its only legislative check would be a Joint Committee on Atomic Energy, itself a unique panel. Unlike any other joint committee then in Congress, it could sponsor legislation. It quickly became a partner with the AEC in nuclear promotion.

"Whereas the fossil-fuel industries were built by private enterprise, the nuclear industry was sired by a government agency and nurtured by a technology created with public funds," explains Richard S. Lewis, long editor of the *Bulletin of Atomic Scientists*. Says Lewis, "At the outset, this was a logical and acceptable consequence of the government's tight control of all aspects of nuclear technology. Industrial atomic energy would come into being only if the government sponsored it."

And those Manhattan Project contractors were not to be left out, by any means. As the act declares. "atomic energy shall . . . be directed toward . . . strengthening free competition in private enterprise."

The act's claim, incidentally, that it would also go "toward improving the public welfare" is the only legal justification for the setting of a new, admittedly profound national direction by federal fiat. It is a legal basis viewed by some as highly questionable under the American political system.

"The public interest was submerged," notes Lewis, to a "joint government-industry power structure with common goals and with access to money and political power." Thus developed the "Atomic Industrial Establishment devoted to the task of building a fission-reactor power economy, eventually to be based on plutonium fuel, that will last for an indefinite future . . . . In this Establishment the AEC functions as banker and engineer and the Congressional Joint Committee on Atomic Energy as *de facto* board of directors . . . . The Atomic Industrial Establishment" was "well on the way toward" trying to shape "the energy future of America."

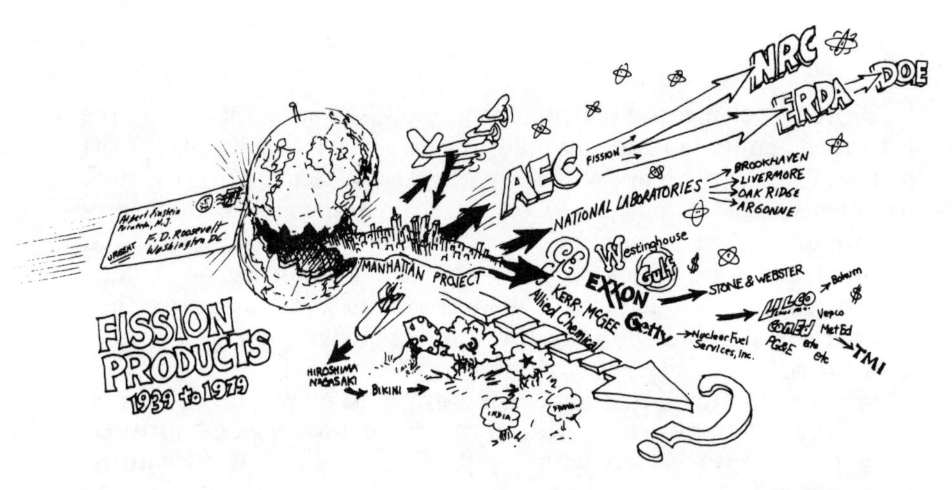

Van Howell

What Lewis calls the "Big Two" of government nuclear contractors became Westinghouse and General Electric—the would-be Coca-Cola and Pepsi-Cola of nuclear power.

They would have what the Brookings Institution calls a "duopoly" on nuclear power. Explains its study: "There can be little doubt that some firms have obtained from their government contract a commercial advantage in their private nuclear business . . . . It is no coincidence that the two monarchs of the civilian nuclear power business, Westinghouse and General Electric, have long operated AEC reactor laboratories . . . . The situation has given rise to the duopoly which exists today in atomic power."

But despite vigorous promotion by the government and pushing by Westinghouse and General Electric to sell the notion of nuclear power plants, the utilities didn't want to have anything to do with them—far too dangerous, they said. Who would pay the damages for a catastrophe? By 1955 no utility had yet agreed to erect a nuclear plant.

So the atomic empire turned to figuring out all other kinds of things to be done with nuclear power—projects paid for, as always, by the taxpayers.

These schemes pursued from the 1950's to the 1970's (some of which are still alive today) include:

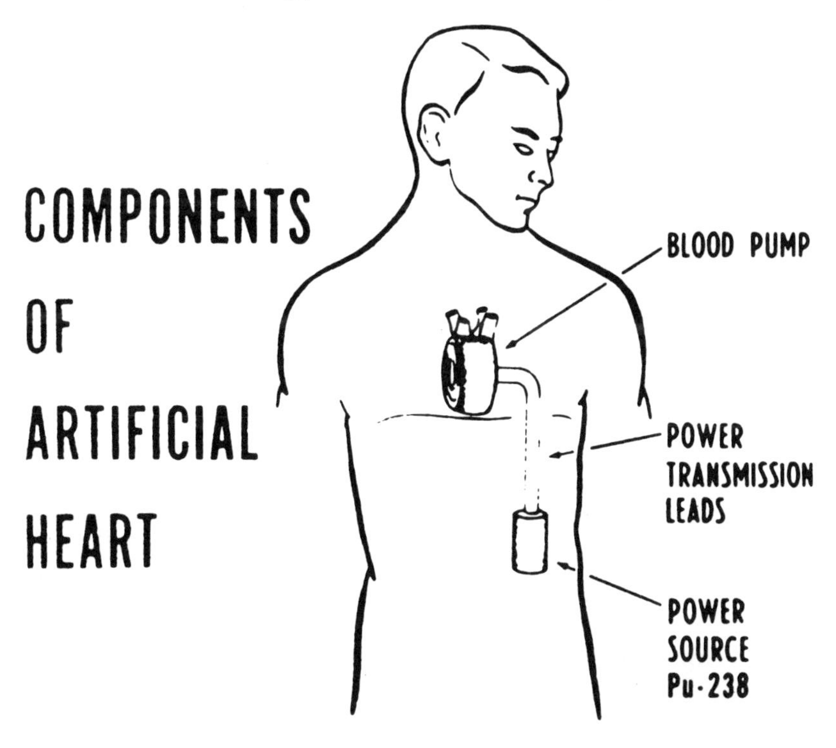

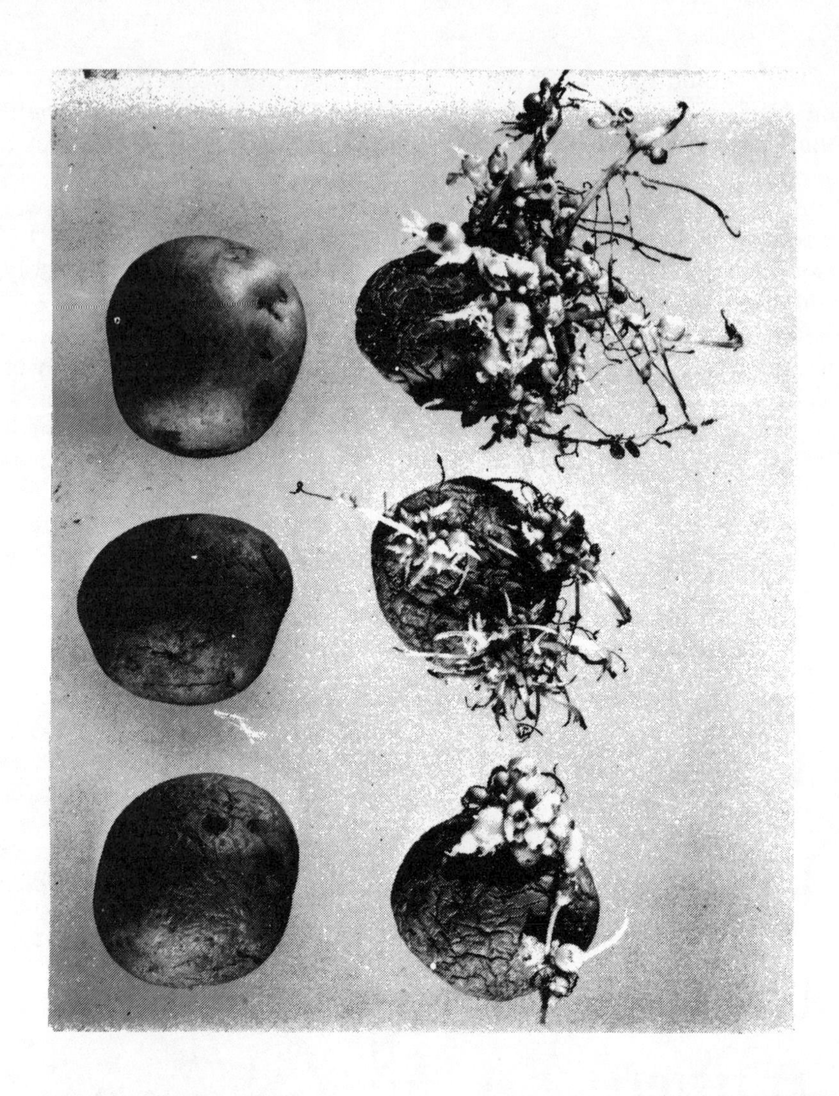

Subjecting food to massive amounts of radiation, up to 50,000 rads (similar to rems), to preserve it. The left row of potatoes was given large doses of radiation. The object, explained Paul McDaniel, the AEC's director of research, is "to achieve optimum balance between the effective time the food can be preserved, the acceptability of the processed food as to color, taste, odor and other characteristics, the general wholesomeness of the food for human consumption, and the overall economic benefits to be derived from the adaption of radiation preservation."

The U.S. government installed a series of what it called "shipboard irradiators" like the one above, each weighing seventeen tons, aboard the above vessel. Crabs were given a dose of 350,000 rads to preserve them and, McDaniel reported, the "flavor and odor of crab irradiated at 350,000 rads were still acceptable after about 40 days." Shrimp went through what is termed "radiation pasteurization" with a dose of 100,000 rads.

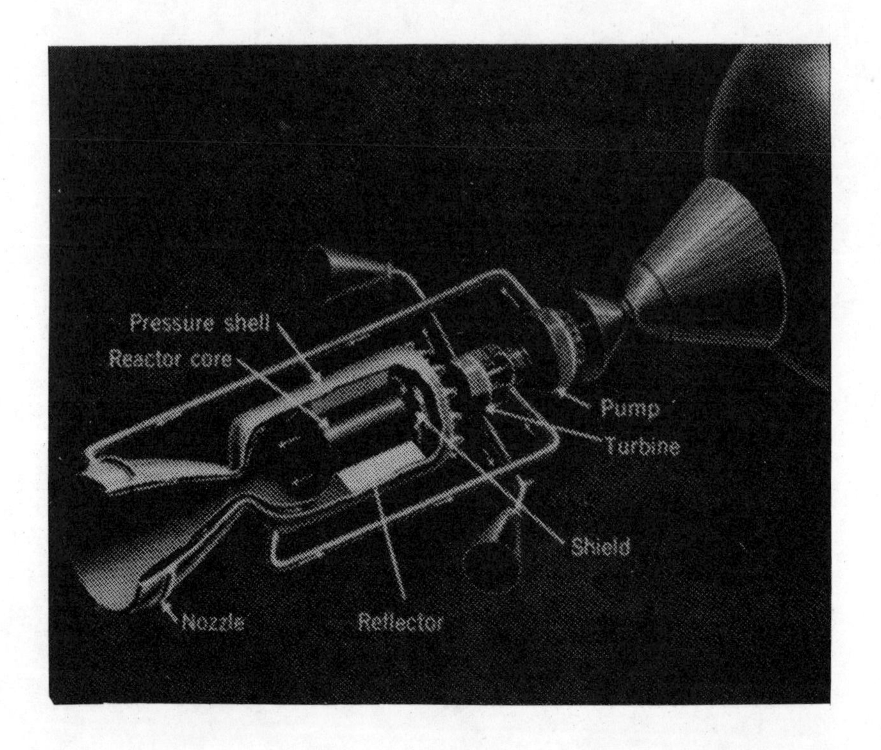

This nuclear-powered rocket engine was built under the NERVA (Nuclear Engine for Rocket Vehicle Application) Program. Concerns about what would happen if a rocket with such an engine crashed finally led to the program's cancellation. Meanwhile, a succession of nuclear-powered satellites was shot into space, like this SNAP-10A. SNAP-9A went out of control and made a flaming re-entry back into the earth's atmosphere, vaporizing its plutonium fuel load into the atmosphere. Several nuclear-powered satellites still are in earth orbit.

XECF

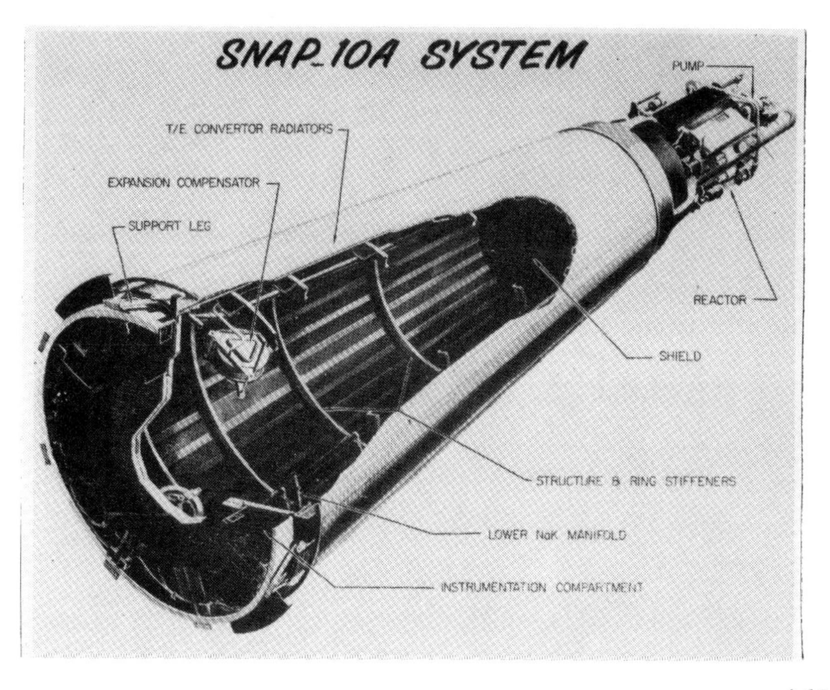
SNAP-10A SYSTEM
PUMP
T/E CONVERTOR RADIATORS
EXPANSION COMPENSATOR
SUPPORT LEG
REACTOR
SHIELD
STRUCTURE & RING STIFFENERS
LOWER NaK MANIFOLD
INSTRUMENTATION COMPARTMENT

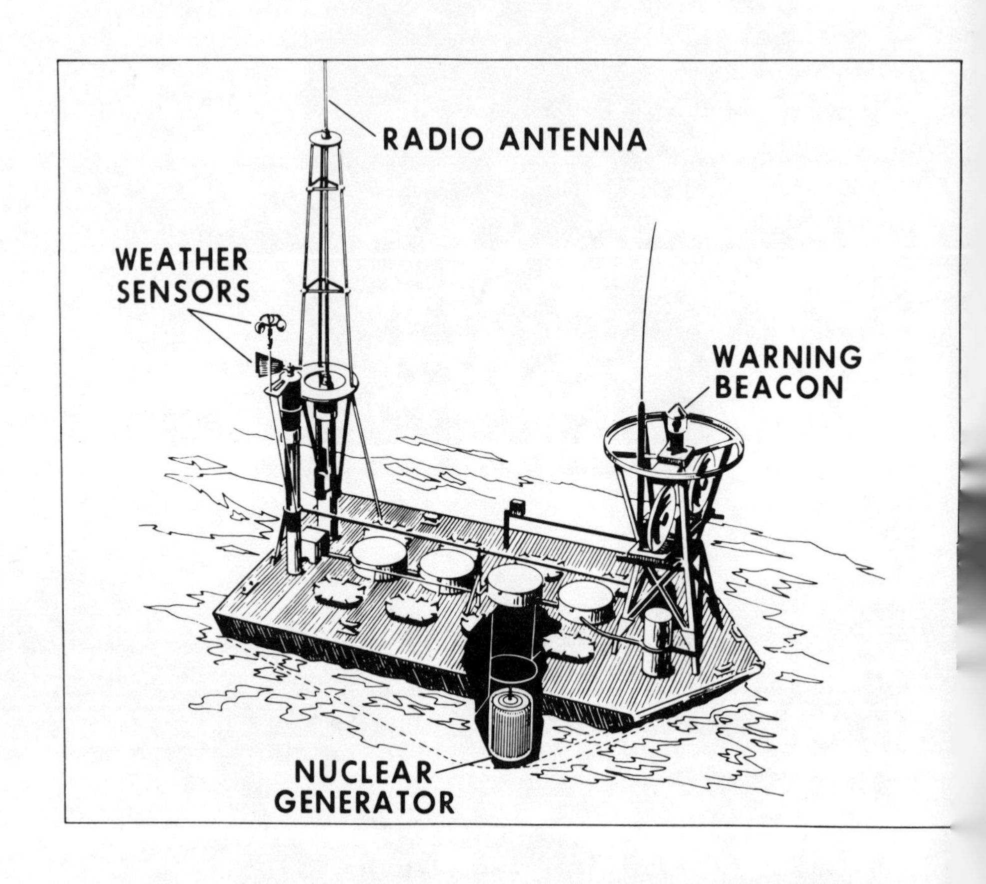

A "nuclear-powered deep sea weather station" was built and anchored in the Gulf of Mexico.

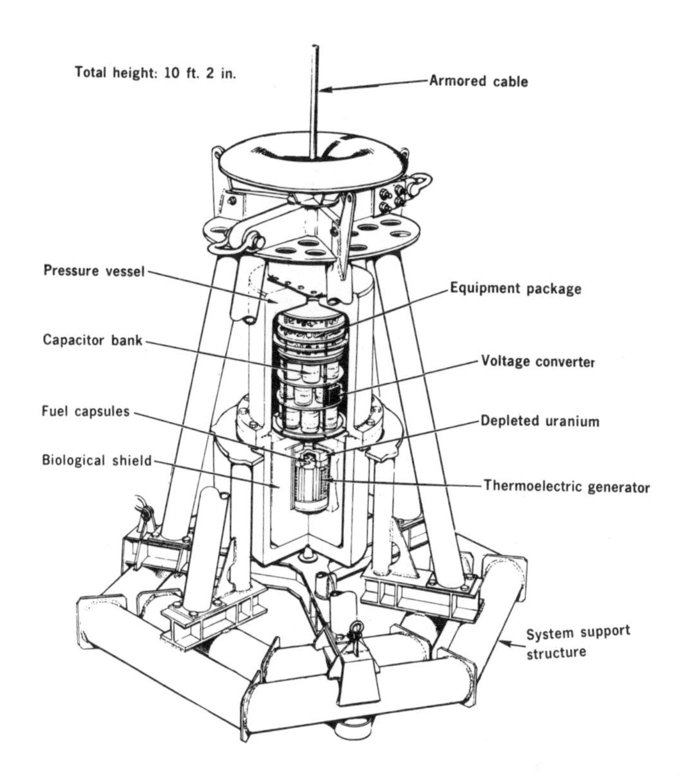

Nuclear-powered light buoys were put in place, this one off Baltimore. These and the "deep sea weather station" used strontium-90 as fuel.

NUCLEAR-POWERED AGRO-INDUSTRIAL COMPLEXES

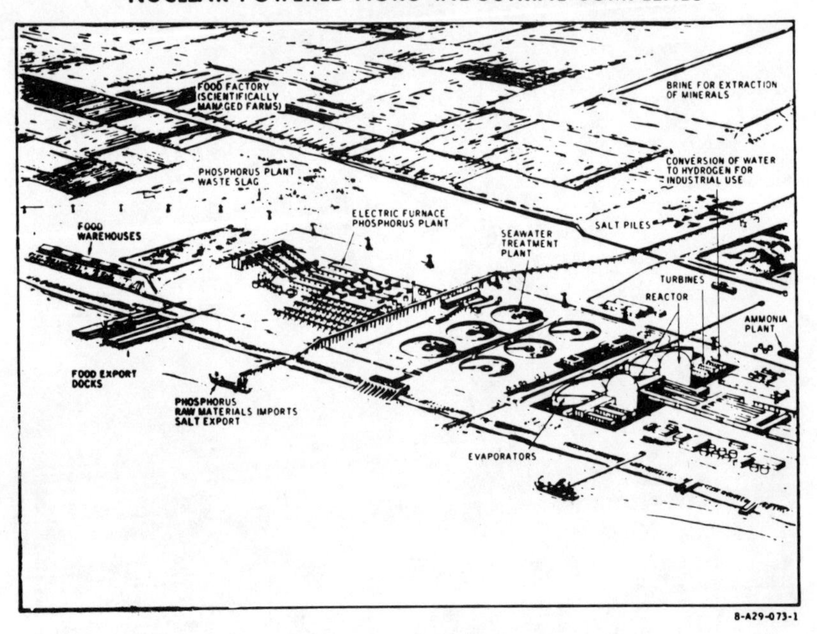

There were plans for "nuclear-powered agro-industrial complexes."

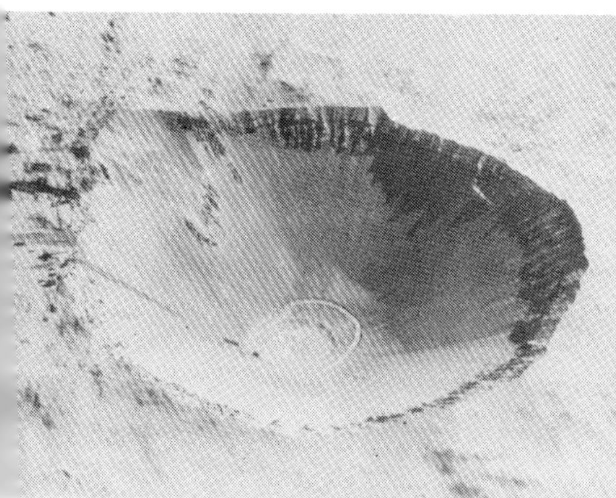
Project SEDAN crater.

Project SEDAN crater — 1216 feet in diameter and 323 feet deep.

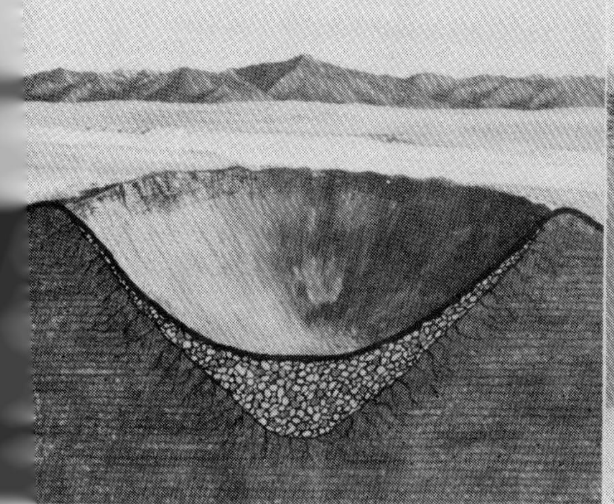
Cross section of SEDAN crater.

A viewing stand at the edge of the SEDAN crater.

And great attention was paid to the use of "nuclear explosives." Explained McDaniel: "Th enormous force of nuclear explosions allows one to 'think big' in planning suitable projects."

Atomic bombs were exploded to see what kinds of holes in the ground they made.

And "Project Plowshare" was organized to make use of "nuclear explosives."

From the AEC's "The Atom and the Ocean":

Project Plowshare

Nuclear explosives are, among other things, large-scale, low-cost excavation devices. In this respect, with the proper pre-detonation study and engineering, they are ideally suited for massive earth-moving and "geological engineering" projects, including the construction of harbors and canals. The western coasts of three continents, Australia, Africa, and South America, are sparsely supplied with good harbors. A number of studies have been undertaken as to the feasibility of using nuclear explosives for digging deepwater harbors. Undoubtedly at some time in the future, these projects will be carried out.

In addition, there are many places in the world where the construction of a sea-level canal would provide shorter and safer routes for ocean shipping, expedite trade and commerce, or open up barren and unpopulated, but mineral-rich lands to settlers and profitable development. The AEC Division of Peaceful Nuclear Explosives operates a continuing program to develop engineering skills for such projects. Construction of a sea-level canal across the Central American isthmus is one well-known proposal for this "Plowshare" program.

The use of nuclear explosives in this manner may one day change the very shape of the world ocean.

For years work continued on plans to blast a canal through the isthmus of Panama with atomic bombs. It was to be called the Panatomic Canal.

Enabling legislation was passed by the U.S. Congress.

AN ACT To provide for an investigation and study to determine a site for the construction of a sea level canal connecting the Atlantic and Pacific Oceans

Be it enacted by the Senate and House of Representatives of the United States of America in Congress assembled, **That the President is authorized to appoint a Commission to be composed of five men from private life, to make a full and complete investigation and study, including necessary on-site surveys, and considering national defense, foreign relations, intercoastal shipping, interoceanic shipping, and such other matters as they may determine to be important, for the purpose of determining the feasibility of, and the most suitable site for, the construction of a sea level canal connecting the Atlantic and Pacific Oceans; the best means of constructing such a canal, whether by conventional or nuclear excavation, and the estimated cost thereof. The President shall designate as Chairman one of the members of the Commission.**

Five sites were chosen and carefully surveyed.
The sites as laid out on an AEC map:

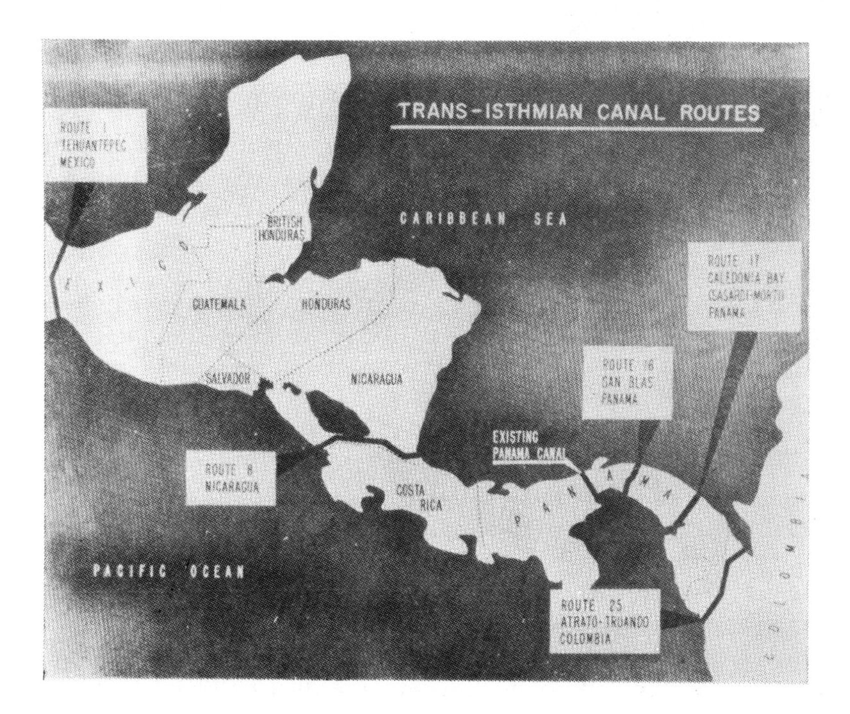

"In theory, at least," explains Lewis, "the Plowshare technologists in the AEC and in the Army knew it would work. What they did not know was the full range of environmental effects."

And they didn't much care.

For one favored alternative, "Route 17," just west of Columbia, "it was estimated that nuclear excavation would require 250 explosives with a total yield of 120 megatons" to create the canal, notes Lewis. "The charges would be detonated in thirty groups, with yields per group ranging from 800 kilotons to 11 megatons."

What about the gigantic amounts of radiation that would be spewed out? Well, the government conducted studies concluding that most of the radioactive debris would be contained in or near the channel, although some would travel. Still, the government insisted, "potential external radiation doses will be far below the lethal levels and well below levels known to produce obvious clinical symptoms in man."

But, asks Lewis, "what about the levels which do not produce 'obvious' clinical symptons in man" but lead to leukemia and cancer later on?

Finally the AEC's Division of Peaceful Nuclear Explosives called off the project because of "prospective host country opposition to nuclear-canal excavation."

As the first chairman of the AEC, David Lilienthal—later highly critical of nuclear development—said: "The classic picture of the scientist as a creative individual, a man obsessed, working alone through the night, a man in a laboratory pushing an idea—this has changed. Now scientists are ranked in platoons. They are organization men. In many cases the independent and humble search for new truths about nature has become confused with the bureaucratic impulse to justify expenditure and see that next year's budget is bigger than last's . . . . Without judging the details of these undertakings, the important thing they show is how far scientists and administrators will go to try to establish a nonmilitary use" for nuclear power.*

Still, beyond the mad scientist schemes of plutonium-powered artificial hearts, zapping food with enormous amounts of radioactivity, nuclear-powered rockets, and using atomic bombs to blast through a "Panatomic Canal," how could the greatest such scheme of them all—using dangerous nuclear power to make electricity—be sold?

The nuclear establishment finally told the reluctant utilities: we'll build nuclear plants if you won't.

"It is the commission's policy," said AEC chairman Lewis Strauss in 1957, "to give industry the first opportunity to undertake the construction of power reactors. However, if industry does not, within a reasonable time, undertake to build types of reactors which are con-

**Change, Hope And The Bomb*, David Lilienthal, Princeton University Press, Princeton, N.J., 1963.

sidered promising, the commission will take steps to build the reactors on its own initiative."

That year, Admiral Rickover and his Navy team completed construction of what would be America's first civilian nuclear plant at Shippingport, near Pittsburgh, using a pressurized water reactor system designed by Westinghouse for naval vessels.

Although a battle had ensued after World War II over who would control nuclear power, it ended in a happy marriage: an ostensibly civilian set-up with the AEC in charge, but with the military deeply involved. Atomic bombs were being built by the gross. By 1952 some thirteen reactors had been constructed to turn out fuel for bombs. Great attention was being paid to nuclear propulsion of warships, particularly submarines, to reduce the need for re-fueling. Some one billion dollars alone went into the development of a nuclear-propelled military airplane until it was scrapped for the same reason the nuclear rocket was dumped: what happens if it crashes? Under civilian cover, the military was able to justify continued large appropriations for nuclear weaponry. Shared facilities were developed.

The biggest breakthrough, also in 1957, came with the passage of the Price-Anderson Act, which limited liability to a utility in the event of a nuclear plant accident to $60 million, a fraction of the expected cost. As a co-sponsor of the act, Representative Melvin Price of Illinois said, "the power development program was bogged down." The insurance industry wouldn't cover nuclear power plants, he noted, and "it wasn't until passage" of the act "that the program got off the ground and they started to build plants." The utilities just "wouldn't risk going into this uncharted area" and paying for the potential damages.

At the same time, the American oil industry–dominated by the Rockefeller family from the beginning–moved into the nuclear power business.

This had something to do with the oil industry seeking control of all possible competititon. It had long been buying up sources of coal and natural gas. But nuclear power in the 1950's was a highly questionable economic risk–as it is, indeed, now. Another factor was the human one. Nelson Rockefeller, a leading member of the Rockefeller family, was attacted to nuclear power; he liked the potential power of it. As to dangers, one associate said his attitude was: "You have to live dangerously."

Nelson Rockefeller "was to nuclear energy what his grandfather, John D. Rockefeller, had been to oil," explains nuclear writer Anna Mayo. The latter day Rockefeller empire would follow many traditions originally laid down by John D.–bringing what it had done in the oil business to nuclear.

Anthony Sampson in *The Seven Sisters*, a history of the major oil companies, speaks of "the secrecy" and "ruthlessness" of John D. "Rockefeller's methods . . . . From his headquarters at 26 Broadway in New York, Rockefeller controlled a corporation unique in the world's history. It was almost untouchable by the state governments which seemed small beside it, or by the federal government in Washington whose regulatory powers were still minimal. By bribes and bargains it established 'friends' in each legislature, and teams of lawyers were ready to defend its positions. Its income was greater than that of most states"; as now, with sales in the tens of billions of dollars, the Standard empire has more income than most countries.

The name of its game has been umbrella monopoly. And the Rockefeller trust has operated by any means. Writes Ferdinand Lundberg in *The Rockefeller Syndrome:* "Standard Oil illegally took rebates and drawbacks, thereby profiting at a great rate, and paid out kickbacks. But it went beyond this: it secretly established ownership in presumably competing companies. And when such ownership became known, it secretly established or bought 'independent' companies in order to deceive the growing number of persons who did not wish to do business with Standard Oil. With its secretly controlled 'independents,' it waged phony price wars, driving true independents to the wall. Then it raised prices."

"The Standard Oil Trust . . . organized for a purpose contrary to the policy of our laws," the Ohio Supreme Court ruled in an early attempt to break up the Standard Trust in 1892; it has as "its object to establish virtual monopoly of the business of producing petroleum, and of manufacturing, refining and dealing in it and all of its products, throughout the entire country, and by which it might not merely control the production, but the price, at its pleasure."

After the work of Ida Tarbell, who wrote the classic book on the Rockefeller oil empire, *The History of The Standard Oil Trust,* and another turn-of-the-century American muckraking journalist, Henry Demarest Lloyd, who wrote *Wealth Against Commonwealth,* the evidence was abundant that through ruthlessness, corruption of government and violation of anti-trust laws, Standard had become a monopoly dominating oil in America–from the wellhead to the pump–and was going world-wide. In 1911 the U.S. Supreme Court ordered a "trust-bust" of Standard Oil for its being a monopoly which functioned "to drive others from the field and exclude them from their right to trade." The Supreme Court ordered Standard broken up into thirty-eight units, most of the main names you see on gas station signs today; Standard Oil of New Jersey known as Exxon or Esso, Standard Oil of New York which became Mobil, Standard Oil of Indiana which goes by Amoco, Standard Oil of California which is Chevron—and so on.

But all the Standard Trust elements still function closely together. The trust, in fact, has never been broken.

As this 1976 Congressional report concludes:

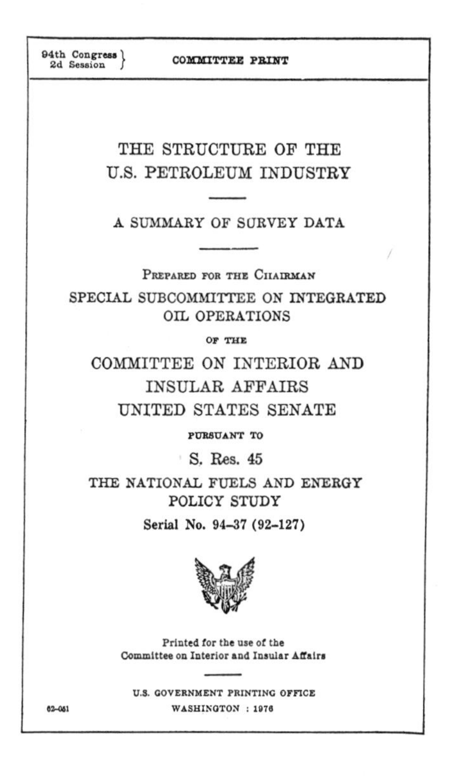

94th Congress
2d Session

COMMITTEE PRINT

THE STRUCTURE OF THE
U.S. PETROLEUM INDUSTRY

A SUMMARY OF SURVEY DATA

PREPARED FOR THE CHAIRMAN
SPECIAL SUBCOMMITTEE ON INTEGRATED
OIL OPERATIONS
OF THE
COMMITTEE ON INTERIOR AND
INSULAR AFFAIRS
UNITED STATES SENATE
PURSUANT TO
S. Res. 45
THE NATIONAL FUELS AND ENERGY
POLICY STUDY
Serial No. 94-37 (92-127)

Printed for the use of the
Committee on Interior and Insular Affairs

U.S. GOVERNMENT PRINTING OFFICE
62-061 WASHINGTON : 1976

(4) The pattern of leading companies in gasoline marketing resembles the marketing territories of the antecedent Standard Oil Trust Companies of 1911.

Big Oil has "myriad relationships . . . based on concentration of control, interlocking directors, financial services, joint ventures, professional conformity, reciprocal favors, commonality of interest . . .

and at its worst, greed and arrogance," explains a 1973 study, *The American Oil Industry, A Failure of Anti-Trust Policy* by Stanley H. Ruttenberg and Associates of Washington. It has used "its vast economic power to influence men and whole nations."

As a 1975 Ruttenberg study, *The Energy Cartel, Big Oil vs. The Public Interest* put it, the oil industry " has learned to create a complex web of official and clandestine relationships that tie official government policy to its own domestic and foreign objectives.... The oil industry is allowed to disregard anti-trust laws." There is "the placement of oil industry officials in high government positions, monumental campaign contributions, some of them illegal ... a vast, highly structured and well-financed lobbying group .... A common description of the oil industry is four integrated sub-industries from extraction at the wellhead to transportation, refining and marketing. This is an obsolete early 20th Century characterization. Today there is a fifth level of integration—the federal government."

Early on, the Rockefellers became leaders of American and global finance. John D. Rockefeller had bridled at the bankers when he began to build his empire, but in the end his family wound up in command of Chase Manhattan Bank and the First National City Bank of New York (Citibank). These banks are bound to many major corporations through stock ownership and interlocking board directorates. Most important, in the development of nuclear power, Chase Manhattan and First National City Bank gained interlocking relations at the director level with Westinghouse and General Electric.

Public relations has also been a hallmark Rockefeller concept. The inventor of this notion of trying to manipulate the facts getting to the people and mold public opinion was Ivy Lee, who became the long-time PR person for the Standard Oil Trust.

The legion of Ivy Lee's PR successors in the oil industry easily blended in with PR personnel of the Manhattan Project who became the AEC information-manipulators and public opinion-shapers.

Another Standard Trust tradition is the industrial strike—the manipulation of supplies to drive up prices or to create pressure for some other goal. By not only controlling the sources but the overall market, a monopoly has blanket marketing power—something you might keep in mind the next time you are waiting on a gas line.

And the oil industry, with all its traditions of monopoly, ruthlessness, corruption, PR, and market manipulation has been merging with the nuclear business.

As the Congressional committee report on "The Structure of the U.S. Petroleum Industry" noted, by 1976:

> Ninety-five percent of uranium milling capacity is affiliated with companies in the oil industry, and the remaining 5 percent is owned by companies with coal properties.

More than half of uranium resources are now owned by oil companies and acquisition continues at a rapid rate. Five of the eight largest uranium mining companies are oil companies: Exxon, Continental Oil, Gulf, Getty and Kerr-McGee. Exxon mines, mills and fabricates fuel rods and is seeking to build a nuclear fuel reprocessing plant. Continental Oil is a major miller of uranium. Gulf Oil has set up General Atomic as a subsidiary and it mills and fabricates nuclear fuel and has been seeking to manufacture nuclear power plants and to go into reprocessing. Getty mills and owns the Nuclear Fuel Services reprocessing facility at West Valley. Kerr-McGee mines and mills uranium and runs uranium and plutonium fuel fabrication facilities, including the Oklahoma plant where Karen Silkwood worked. Atlantic Richfield has operated the nuclear waste storage facilities in Hanford and owns Anaconda Company, a major miner of uranium. And the list goes on.

In 1970, U.S. Senator Albert Gore said, "I can see a situation not far off when we . . . may well have all major energy sources—petroleum, coal, uranium—under the control of a very few powerful corporations."

"The situation with the uranium and nuclear industry," explains *The American Oil Industry* study, is that "the oil industry is rapidly acquiring the production, reserves and milling capacity . . . to control uranium, the raw material upon which the electric utility industry would depend. With the single-mindedness of a hawk sweeping towards its target, the oil industry through its accelerated acquisition of uranium segments, now holds a sharply honed scalpel against the electric industry jugular."

With the oil industry now central to it, the nuclear industry has become tightly concentrated and interlocked, too.

The Arthur D. Little Company, a management analysis firm, made a study for the U.S. government on the matter in 1968 (to which the government paid little attention).

NYO-3853-1
TID UC-2

COMPETITION IN THE NUCLEAR POWER SUPPLY INDUSTRY

Report to

U.S. Atomic Energy Commission
U.S. Department of Justice

prepared by

Arthur D. Little, Inc.

Contract No. AT (30-1)-3853

December 1968

The study stressed:

Competition between nuclear and fossil as a factor of significance in the energy market is dependent upon the existence of separate companies in the two fields. If the companies serving the two markets were to become the same by evolving into nuclear/fossil energy companies, the nature of competition between them would change significantly. The development of firms diversified in the energy field could pose a threat to inter-modal energy competition. Whether this in turn would impair competition in the overall energy market would depend upon the number of firms and levels of concentration in the overall market.

The analysis speaks of concentration in the building of nuclear plants, described as NSSS (nuclear steam supply systems).

In addition to the high degree of concentration in major segments of the nuclear power supply industry, there is substantial vertical integration. Each of the four NSSS suppliers is extensively vertically integrated in NSSS components, although these companies vary significantly with regard to the products in which integration occurs. For instance, of the four NSSS suppliers, only B & W and CE manufacture pressure vessels, and only GE and Westinghouse produce turbine-generators. The four NSSS suppliers are vertically integrated in nuclear fuel fabrication, and a number of companies in uranium mining and milling are either integrated to various degrees in the fuel cycle (Kerr-McGee and United Nuclear) or intend to expand in the fuel cycle (Gulf General Atomic, Getty Oil and Atlantic Richfield).

(B&W and CE stand for Babcock and Wilcox and Combustion Engineering, two minor plant manufacturers after General Electric and Westinghouse.)

Interlocking directorates were found to be widespread.

The interlocking directorate is a form of integration sometimes considered beneficial from a managerial point of view but sometimes thought to be potentially injurious to competition. We have made a detailed (though not exhaustive) examination of the corporate interlocks among the primary nuclear reactor/fuel suppliers and electric utilities, the results of which appear in an appendix to Part Two of this report.

Here is the chart assembled for the government's study:

From : NYO 3853-1 TID UC 2

"COMPETITION IN THE NUCLEAR POWER SUPPLY INDUSTRY"

Report to US Atomic Energy Commission
US Department of Justice

Prepared by Arthur D. Little, Inc. Contract No. AT (30-1)-3853
December 1968

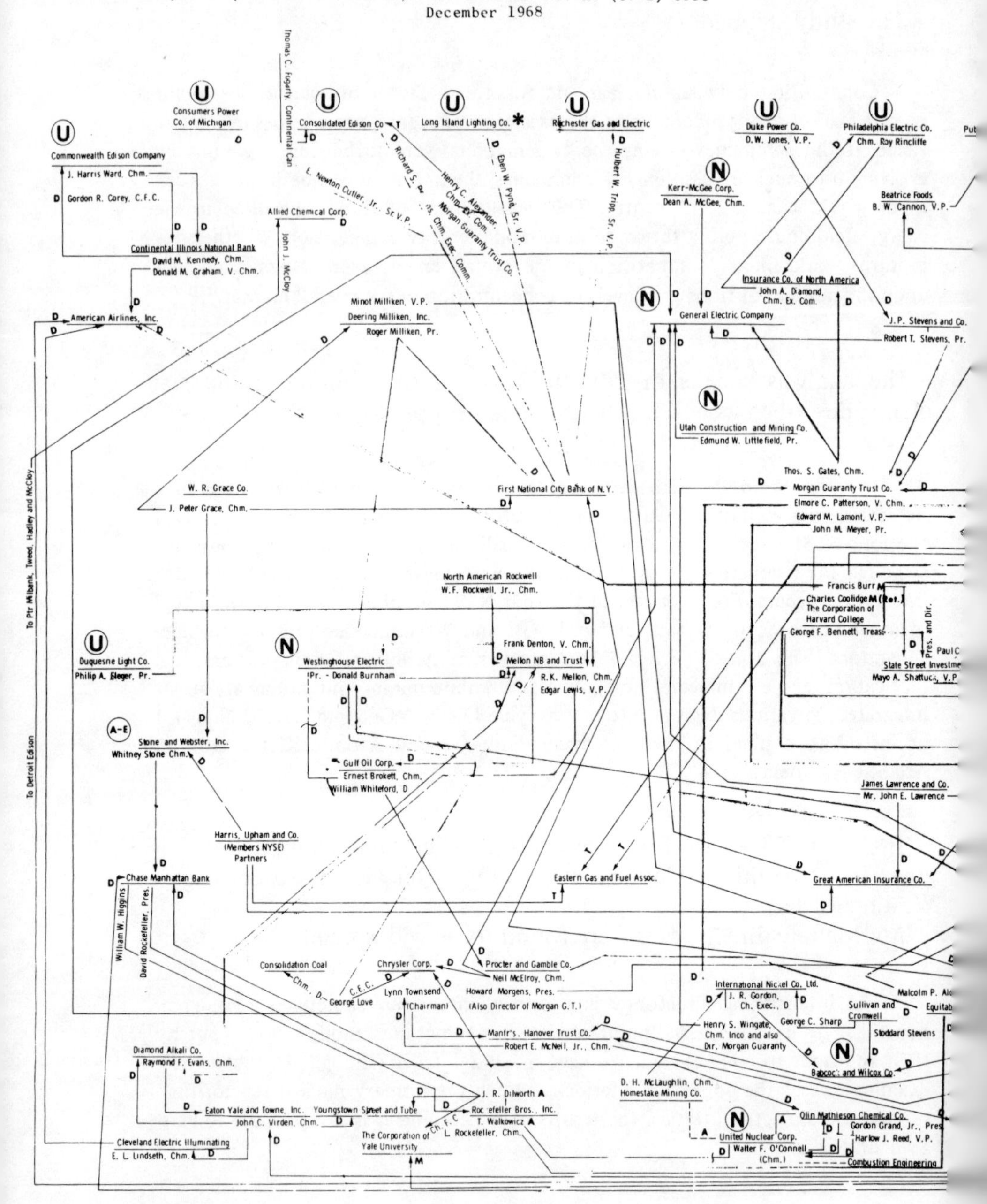

Sources: Moody's Handbook of Corporate Management 1967
Company proxy statements, and annual reports
Standard and Poor's Register of Corporations Directors, and Executives, 1968.

FIGURE B-1 CORPORATE RELATIONSHIPS THROUGH DIRECTOR AFFILIAT

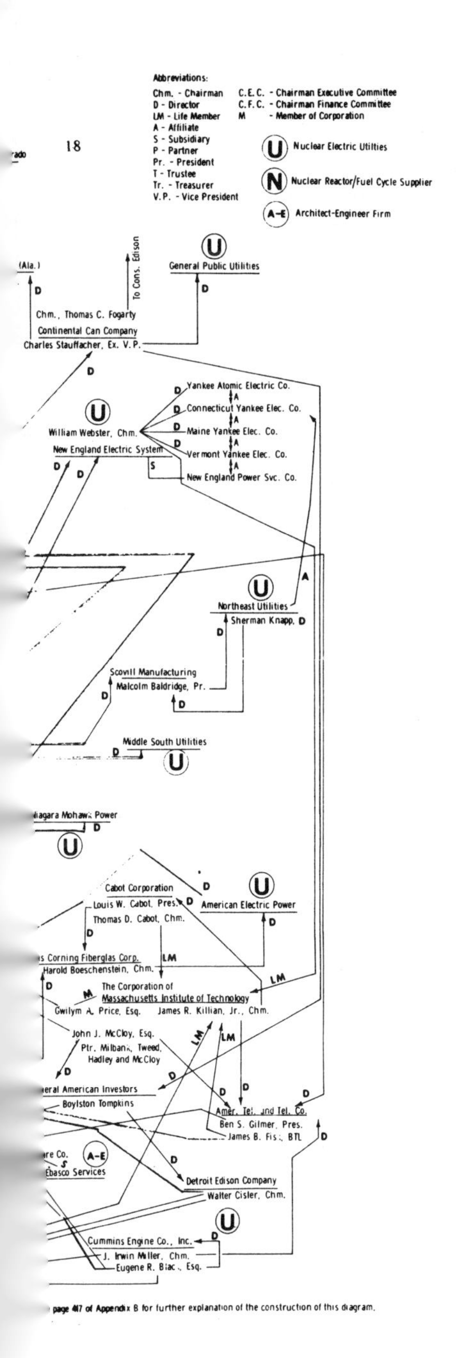

*For a sample interlock, note the name of Eban W. Pyne under the Long Island Lighting Company on the chart. He sits on the board of LILCO (one of the most nuclear power-oriented utilities in the nation) with, among others, Phyllis S. Vineyard, wife of George Vineyard, director of Brookhaven National Laboratory and William J. Casey, the campaign director for the presidential race of Ronald Reagan. Pyne, who owns 7,000 shares of LILCO stock–worth over $100,000–is also senior vice president of Citibank, called at the time this investigation was made, First National City Bank of New York. The bank not only interlocks on the director level with other utilities: Con Edison, Rochester Gas & Electric; but as the arrows note, into corporations in the nuclear supply business. Meanwhile, LILCO has multi-million dollar loans with Citibank. Pyne, in charge of the bank's Long Island operations, joined the LILCO board in 1959 after LILCO received its first $12 million loan from his bank. LILCO had been going through a corporate reorganization and was considered a questionable financial risk. No other bank would lend it major amounts of money. Citibank has, meanwhile, purchased substantial LILCO stock and serves as transfer agent for LILCO stock. Pyne is also the Citibank link wtth W.R. Grace & Co. as a board member of that chemical and oil conglomerate shown on the chart. For many years Whitney Stone, chairman of the nuclear engineering company of Stone & Webster until his death in 1979, was also a member of the board of W.R. Grace & Co. Stone & Webster has close ties with LILCO. Even though it was the higher bidder among four engineering firms, Stone & Webster got the contract to be the "architect-engineer" of what is supposed to be LILCO's first nuclear plant, built at Shoreham, Long Island, N.Y. In 1980 the utility announced the projected cost of the plant had risen to $2.2 billion, ten times the $271 million it was supposed to cost, making it the most expensive nuclear power plant in the world. Stone & Webster has a substantial interest in the Transcontinental Gas Pipe Line Corp. which provides LILCO with most of its natural gas. Stone was also a member of the board of Chase Manhattan Bank, another source of multimillion dollar loans for LILCO. Furthermore, his firm constructed a nuclear reactor at Brookhaven National Laboratory.

The analysis went on:

The data in Table 9 show that a number of major oil companies, by entering other energy-producing industries, are becoming full-line energy companies. Eleven of the top 20 oil companies are now in uranium exploration or mining, and five are in coal mining. Five of these diversified energy companies are among the top ten firms on *Fortune's* list of the 500 largest U. S. industrials.

TABLE 9

DIVERSIFICATION BY MAJOR PETROLEUM COMPANIES IN ENERGY-PRODUCING INDUSTRIES
(X means that company is engaged in activity)

Company	Assets in 1965 (in thousands)	Rank in assets		Crude oil production		Petroleum refining		Natural gas	Oil shale[1]	Coal	Uranium
		Petroleum industry	Fortune's 500 largest industrials	Domestic	Foreign	Domestic	Foreign				
Standard Oil (New Jersey)	$13,073,437	1	1	X	X	X	X	X	X		X
Texaco	5,342,903	2	5	X	X	X	X	X	X	X[2]	X
Mobil Oil	5,212,380	3	6	X	X	X	X	X	X		X
Gulf Oil	5,210,833	4	7	X	X	X	X	X	X	X	X
Standard Oil of California	4,165,825	5	9	X	X	X	X	X	X		
Standard Oil (Indiana)	3,514,102	6	12	X	X	X	X	X	X		X
Shell Oil	2,671,464	7	15	X	X	X	(3)	(3)	X		X
Phillips Petroleum	2,029,064	8	20	X	X	X	X	X	X		X
Union Oil of California	1,758,516	9	24	X	X	X	X		X		
Sinclair Oil	1,694,519	10	27	X	X	X	X	X	X		
Continental Oil	1,679,473	11	28	X	X	X	X	X[4]	X	X	X
Cities Service	1,635,417	12	30	X	X	X	X		X		X
Sun Oil	1,205,783	13	41	X	X	X	X	X	X	X[5]	
Tidewater Oil	995,541	14	50	X	X	X	X	X	X		X
Atlantic Refining[6]	993,849	15	51	X	X	X	X	X[4]	X	X	X
Marathon Oil	785,832	16	72	X	X	X	X	X[4]			
Sunray DX Oil	639,204	17	82	X	X	X	X				
Standard Oil (Ohio)	617,955	18	88	X	X	X	X		X		
Signal Oil & Gas	597,047	19	96	X	X	X	X	X			
Skelly Oil	420,339	20	128	X	X	X	X				X

[1]Ownership of oil shale lands, exploration rights, or research.
[2]In Western Germany.
[3]By parent company.
[4]Part ownership of refineries in foreign countries.
[5]Ownership of coal-producing lands.
[6]In 1966, Atlantic Refining acquired Richfield Oil and changed its name to Atlantic Richfield. The combined assets of 2 companies in 1965 were $1,488,319,000.

Source: Hearings before the U. S. Senate Subcommittee on Antitrust and Monopoly. Part 1. April 18, 19, 20, 25, 27; May 2, 5, 1967; p. 618

The press—which as in the muckraking days of Ida Tarbell could act as a watchdog, alerting the public to abuses of industry and government, has not (with few exceptions) reported fully on nuclear power through the years. This is because of the intense public relations drive by the nuclear establishment to manipulate, intimidate and silence it. That a major part of the press in recent decades has interlocked with the concentrated U.S. corporate structure is a factor, too. The Atomic Industrial Forum, the lobbying arm of the nuclear establishment, lays out media manipulation strategy in a 1974 internal memo:

The national media, with the middleman of the reporter and editor, cannot be relied on to publish a full and balanced account of nuclear power. The economic benefits of nuclear power will be the most important element of this story, with its ability to help slow the inflationary spiral which is now being encouraged by steady increases in petroleum and coal costs.

The targets for these activities will be not only the industry and the media, but such other influential groups as these (in order of priority):

1. Governmental Decision-Makers (Congress and its staff, federal executive agencies, state legislators and Public Utility Commissions)

2. Influential Organizations (emphasizing the financial community, labor unions, the education community, and major civic groups)

3. Other Interested Segments of the Public (including major business or professional associations, industry employees, etc.)

The memo suggests:

--Press conferences, using every opportunity available to get AIF officials, scientific experts and others before the national media, especially in Washington

--Media field visits, directly informing influential representatives of the national media first-hand about nuclear power and the fuel cycle

--Radio "actualities", through which developments important to the industry could reach radio stations directly

--News feature placement, to "ghostwrite" and place positive articles on behalf of distinguished experts, to place feature stories through syndicate services, and to initiate and place special stories on a near-exclusive basis for prominent publications and special interest media.

and tells how the AIF will "stage-manage news":

POSITIVE NEWS EVENT GENERATION

Press Conferences

One of the important ways in which AIF can step in to help fill the information gap that is being created by the demise of AEC and of much of the Congressional nuclear leadership is to stage-manage news

---

From time to time, PAIP has staged luncheon briefings for newsmakers and these, too, should be expanded. Allied to this effort should be an aggressive campaign to get our senior spokesmen on radio and TV panels and interview shows.

Media Field Visits

At least once a year, and preferably more than that, PAIP should organize and conduct a tour for top-level journalists of one or more aspects of the nuclear fuel cycle. For example, in 1975 a most valuable tour could take this select group through the entire front end of the fuel cycle, from mine to mill to enrichment, fabrication and core loading. Another tour, probably not realizable before 1976, could be geared to the LMFBR.

Press junkets are tricky, but enormously worthwhile, not so much for the spot coverage they generate but for the lasting feel for a complex story that they give a writer. Travel expenses, ideally, should be borne by the guests, but even so the logistic and amenity expense can be fierce.

PROPOSED ADDITIONS TO AIF
PUBLIC AFFAIRS AND INFORMATION PROGRAM page 6--

Radio "Actualities"

Several outside organizations offer an "actuality" service: they can record and edit for broadcast statements and comments by AIF officials and handle distribution of these beeper feeds to selected radio markets. A beeper radio system would be extremely useful for quick general response by an AIF spokesman to statements by critics; to propagate an AIF policy statement or urgent press release or simply to offer capsule features on nuclear industry milestones. Surprising as it may be, hund of radio stations will air taped messages coming from outside sources. Professional services can pinpoint distribution to geographical areas o greatest interest to AIF.

News Feature Placement

PAIP's traditional media approach is to mass-distribute news releases, backgrounders and INFO. More and more, however, we find that selected placement, on an exclusive or near-exclusive, basis has merit, and we are doing this within our limited resources. We have also experimented with outside mat services, which blanket small papers. Benefits versus cost for such a service remain in doubt.

The time is ripe for the PAIP staff to ghost-write substantive features for well-known experts and to place these with major consumer publications. At the same time, we should aggressively seek space in these publications for unusual pro-nuclear articles.

CONTACT WITH INFLUENTIAL ORGANIZATIONS

Such influential segments of the public as the financial community, labor unions and teachers would be included in the direct contact and mailings as described in the Nuclear Policy Information Service. Visits, briefings and seminars would be coordinated by the staff, to include AIF member representatives and other experts in nuclear power. Specific goals would be to place articles in their publications, arrange speakers for their conferences and generally encourage their vocal support of nuclear power.

FIELD SERVICE TO MEMBERS

Over the years, the PAIP staff has frequently been asked to consult with members on information projects, publications, media relations, etc. These services have taken various shapes, including telephone consultation, briefings by two or three PAIP staffers for a member company executive staff and full-scale management briefings utilizing outside speakers. As pressure from critics increases, the ability of the AIF to provide such direct information and advice on a rapid-response basis, must also increase to help its members cope with local variations of the national problem.

More and more frequently, local events are developing industry-wide significance--moratorium efforts, state and local legislative hearings--and as often, they are orchestrated by broad-based opposition groups on a common theme. There is a need to provide professional troubleshooters who can spend a few days or a week on location advising local members of other experience, working with local media and opinion leaders, and making known what other Forum and Industry resources can be brought to bear.

THE BREEDER

An issue of special importance in the near future is the breeder reactor. Go or no-go decisions to be made on the breeder in the coming months may well have direct impact on all nuclear power, especially as critics relate the common issues of concern: plutonium, wastes, economics, safety, need for power, etc. There is currently a vacuum in handling breeder public information on a national scale, in terms of working with the national media, with key opinion makers and with the national nuclear industry. For the benefit of all the industry, the Forum should fill this vacuum by establishing a generic national breeder information program which will serve as the principal point of contact with the media, government agencies and other influential groups. A Breeder Information Project Director will be located in Washington and, calling on the other Forum staff, will coordinate generic breeder information, including such peripheral issues as safeguards and plutonium, and provide the necessary contact, publications and a national press seminar.

SPEAKERS BUREAU

There is an increasing demand for a quality fast-response cadre of speakers on nuclear power subjects who can be used for important debates, television talk shows, conferences and hearings. The Forum must expand its speakers bureau, in terms of the number of available speakers, and upgrade their effectiveness by providing them with continual timely information and critique of their performances. To turn this effort into a fully professional one, however, will require two additional commitments:

1. Training of speakers. There are now only a handful of skilled, knowledgeable speakers who can cope with debates, hostile audiences and skeptical interviewers. Many other knowledgeable professors and consultants can be trained for such roles, to the benefit of the entire nuclear industry. The Forum will select a half-dozen candidates, educate them on the issues, and train them on speaking and confrontation tactics to make them more effective as speakers.

2. Speakers on short notice. To provide greater assurance that first-rate speakers would be available for important meetings on short notice, the Forum would place three expert speakers on retainer, these speakers would be committed to providing limited service on virtually any kind of advance notice. AIF member companies with meetings important enough to warrant this service would reimburse the Forum, plus expenses of the speaker. The retainer would also call for the speaker to help the Forum in less time-consuming ways, such as reviewing manuscripts and writing letters to editors of key publicatior

The officers of the Atomic Industrial Forum at the time of the memo are listed as:

OFFICERS

Chairman and Chief Executive Officer:

John W. Simpson, Director-Officer, and Chairman of The Energy Committee
Westinghouse Electric Corporation

Vice Chairman:

Douglas M. Johnson, President
United Nuclear Corporation

Clyde A. Lilly, Jr., President
Southern Services, Inc.

George J. Stathakis, Vice President and General Manager
General Electric Company

President and Chief Operating Officer:

Carl Walske
Atomic Industrial Forum, Inc.

Executive Vice President and Secretary:

Edwin A. Wiggin
Atomic Industrial Forum, Inc.

Vice Presidents:

George L. Gleason
Atomic Industrial Forum, Inc.

Paul Turner
Atomic Industrial Forum, Inc.

Treasurer:

Thomas H. O'Brien, Vice President
First National City Bank

Examples of media manipulation on nuclear power abound.

One example: Emmy Award-winning California filmmaker Don Widener speaks of being woken in the middle of the night by the Atomic Energy Commission's public relations director who threatened: "You going to use anything on plutonium in your movie? Well, don't even mention it. If you do, we'll go to higher sources to stop it." Widener was putting together a documentary for TV, "The Powers That Be." The nuclear establishment protested loudly to the Public Broadcasting Service, which had become dependent on grants from Exxon, Mobil, etc. (it's sometimes described as the Petroleum Broadcasting Service) on this film, and it protested even more when Widener made "Plutonium: Element of Risk." After the industry barrage of complaints on that film, only twelve of the 268 PBS affiliates around the nation would broadcast it. All of this was part of an "ongoing effort to suppress" coverage on nuclear power, says Widener.

Another: "Danger: Radioactive Waste" was a straightforward NBC News documentary on the nuclear waste problem. It was aired but once (on a night the popular "Roots" was on). Don't expect to see it again. The Atomic Industrial Forum organized a massive protest campaign directed at the top executives of NBC, the program's sponsors and federal officials with communications field powers, while Westinghouse had its employees flood NBC with complaints.

Still another: Filmmakers Buzz Hirsch and Larry Cano attempted to make a movie about Karen Silkwood, killed on her way to show a reporter evidence she had compiled for her union on the Kerr-McGee plutonium plant in Oklahoma–safety violations, incidents of workers being contaminated, and radioactive releases. Previously her apartment had been broken into and dusted with plutonium, much of it put on food in her refrigerator. The manila folder of evidence which eyewitnesses had seen her bring with her was missing from her car, which had been run off the road. Hirsch and Cano, while working on the film, were investigated by security personnel of Kerr-McGee. The house of an assistant where they were keeping files and papers on the project was broken into. After the break-in, a police detective's first question, they recall, was "How deep are you in the Silkwood case?" Then Kerr-McGee brought a lawsuit to obtain all their research materials for the film. Warner Brothers suddenly cooled on the film.

Consistently, the nuclear safety story has won a place on the annual list of "ten best censored stories" compiled as a media research study at Sonoma State College in California, with nationally prominent journalists and writers on the panel doing the judging. Columnist and former presidential press secretary Jerry ter Horst cites "media dereliction, neglect and lack of perception" for this.

"I built a full-time career on covering nuclear horror stories that *The New York Times* neglected," says Anna Mayo, one of the few journalists in the nation whose work on nuclear power is published (in *The Village Voice* in New York).

"One of the clearest examples" of press resistance to probe into nuclear power, said Ms. Mayo, "is *The Washington Post*," long-owned by Eugene Meyer "who was also a founder of Allied Chemical," which has been deep in the nuclear business. His daughter is Katharine Graham, chief executive officer and chairman of the board of the Washington Post Company which owns *The Washington Post, Newsweek* and a string of television stations. She has also been a member of the board of Allied Chemical. Allied Chemical runs a major nuclear fuel plant in Metropolis, Illinois, for many years ran a nuclear reprocessing plant in Idaho for the government, and is partners with General Atomic Co. in the Nuclear Fuels Reprocessing Center in Barnwell, S.C.

"When the *Post* reports radiation leaks they always give you the impression that all that is needed is better monitoring and better safety equipment," says Ms. Mayo. She speaks of a *Washington Post* reporter who told her that he "was embarrassed at how he had been prevented from writing in this area."

"We know that they've had the information," Ms. Mayo went on, "but they have not printed many, many stories."

Similarly, the other national U.S. newspaper, *The New York Times,* Ms. Mayo said, has reporters "who are interested in this issue, and well able to report it, and specialized in this area, and have all the resources of the *Times.*" But "they are often reined in, put on other assignments."

A veteran *New York Times* reporter, requesting anonymity, put it this way: "It is understood at the *Times* that the nuclear story is to be soft-pedaled." Why? As this *Times* staffer saw it, "The *Times* is part of the establishment."

After a nuclear industry complaint, said Ms. Mayo, her articles were dropped from the Long Island daily paper, *Newsday.*

An example of what goes on, Ms. Mayo pointed out, "is the story I just did on Dr. Ernest Sternglass' correlations between fetal exposure to radiation and the drop in the SAT scores seventeen, eighteen years later, particularly in states such as Utah.* The story was not covered by the *Times.* And they had a reporter at the conference

*Which received high levels of radioactive fallout after A-bomb tests.

where he presented his paper . . . . As soon as my story came out in the *Voice,* the morning it came out, he was invited by the Today show to make an appearance while he was in New York. Later in the day the Today people called him back and told him they were sorry but they would have no need of him on the program. So he said, 'Is it a time conflict? I would be able to stay in New York longer if that's the problem.' And the spokeswoman for the Today show said, 'No, there will not be any time for you ever to go on the Today program.'"

Alden Whitman, a *Times* reporter for twenty-five years, now retired, said, "There certainly was never any effort made to do" in-depth or investigative reporting on nuclear power. "[David] Burnham tried often enough and never was given any opportunities to do it," said Whitman. "They wouldn't spring him. They wouldn't allow him the manpower or the time or anything else." Why this attitude by the *Times* on nuclear power? "I think there is stupidity involved," said Whitman. Further, "The *Times* does regard itself as part of the establishment . . . . They get very nervous when they attack industry. Certainly when they attack industry that is heavily involved in finance and the banks as nuclear power is, they would get very up tight. They don't want to attack the status quo." Even in the wake of Three Mile Island, said Whitman, the *Times'* stories on nuclear power have been "tucked away, put in the middle of the paper."

Marian Heiskell, a member of the board of The New York Times Company and the sister of Arthur Sulzberger, the *Times* publisher, and wife of Andrew Heiskell, president of Time, Inc. is an active member of the board of Con Edison.

This writer did investigative reporting for the *Long Island Press,* a major daily in the New York area. When the nuclear issue first began developing on Long Island, I was steered away from it—although investigative reporting in the environmental and energy fields had been my specialty. A junior reporter was sent to Buchanan, New York, to write a five-part series on how the people in that town liked the Consolidated Edison Indian Point plants two miles away, the first nuclear plants in the New York area. The reporter was then assigned to cover the licensing hearings for what would be Long Island's first nuclear plant, at Shoreham. The reporter received a $1,000 award for the coverage from the Atomic Industrial Forum. Nevertheless, as the intervenors pressed their case and the reporter covered their arguments, the local utility, the Long Island Lighting Company, brought pressure on the *Press* and the reporter was taken off the nuclear issue, put on full-time arts and theatre coverage and subsequently resigned from the paper.

I first walked into a nuclear story upon encountering a weather station erected on a Long Island beach by the Atomic Energy Commission. What might an AEC weather station be doing down from the well-known "Hot Dog Beach" in the Hamptons? It was put there, my inquiries found, to monitor how a radioactive cloud would hit the shoreline following an accident at a floating nuclear power plant. A seventy-five foot landing craft on loan from the Navy was being used, along with aircraft and a trawler. Huge clouds of white smoke were discharged on the ocean to see where they would float. Those involved in the study said that because of offshore ocean winds, the smoke usually floated to the Long Island shore. Upon finding that I was working on this story, the editor of the *Press* said I should "play down" the story. I pointed out that the existence of the weather station had never been revealed, that it had been doing experiments for years and that no news of floating nuclear plants planned for placement south of Long Island, along New Jersey, had been reported on Long Island.

"We don't want to get people upset," said the editor.

This kind of experience is widespread throughout the media.

"One of the most expensive and well-orchestrated public relations jobs ever has been done with the nuclear power business," explains Geoffrey Cobb Ryan, American editor of *Index on Censorship,* a journal published in England and regarded as the top international publication dealing with media censorship and cover ups. Ryan, who has investigated nuclear power coverage, says that beginning with government, then continuing with industry, the massive nuclear PR effort "snowed the media." He speaks of limits placed because "all publishers like economic progress and a booming economy, and nuclear has been closely linked through PR efforts to economic well-being and progress and development and continuing growth." He describes much of the media as being "connected, the people who control the media, their publishers, their boards of directors" to elements of the nuclear industry, from "the utility firm which services the newspaper's area" to other companies within the nuclear cycle. He notes that the Graham media group is "notoriously optimistic about a nuclear future, reluctant to publish adverse information. They downplay it."

In December 1979 the *Columbia Journalism Review* published the results of a study* of directors of the nation's twenty-five largest

*"Interlocking Directorates," by Peter Dreier, assistant professor of sociology at Tufts University and a former newspaper reporter and Steve Weinberg, former business reporter for *The Des Moines Register,* who directs the University of Missouri School of Journalism Graduate Reporting Program in Washington, D.C.

newspaper companies showing "thousands of interlocks with institutions the papers cover—or fail to cover every day." Newspaper corporate directors who are also directors of companies in the nuclear business include Clark M. Clifford, director Knight-Ridder Newspapers and Phillips Petroleum; Wilmot R. Craig, director of The Gannett Company and Rochester Gas & Electric; James E. Webb, Gannett and Kerr-McGee; Thomas G. Ayers, director of The Tribune Corp. (which publishes *The Chicago Tribune* and *The New York Daily News*) and Commonwealth Edison and Breeder Reactor Corp.; J. Paul Austin, Dow Jones & Company and General Electric and Continental Oil; James Q. Riordan, Dow Jones and Mobil Oil; Richard D. Wood, Dow Jones and Standard Oil of Indiana; Walter B. Gerken, The Times Mirror Company (which publishes *The Los Angeles Times, Newsday, The Dallas Times Herald*) and Southern California Edison; William J. Casey, Capital Cities Communications and the Long Island Lighting Company.

The Media Institute,* in a study of how much time the three TV networks devoted to news about nuclear power on their evening newscasts between 1968 and 1979, found that it came to one quarter of one per cent of the time available for news—an average of less than four seconds for each twenty-two minute newcast.

Westinghouse, as "Group W" (Westinghouse Broadcasting Company), itself owns five television stations (the limit set by the Federal Communications Commission)—WBZ-TV in Boston, WJZ-TV in Baltimore, KYW-TV in Philadelphia, KDKA-TV in Pittsburgh and KPIX-TV in San Francisco, and numerous radio stations including three all-news stations, WINS in New York, KYW in Philadelphia and KFWB in Los Angeles. It also syndicates P.M. Magazine, the nation's major nightly TV news magazine.

General Electric owns three TV stations—WNGE-TV in Nashville, WRGB-TV in Schenectady, KOA-TV in Denver, and several radio stations. G.E. has been involved in arrangements to buy up the TV and radio stations owned by Cox Communications. It would be the largest merger in broadcast history.

Both Westinghouse and G.E. are major sponsors of broadcast programming.

An example of how G.E. will not tolerate anything critical about nuclear power on programming it is connected with occurred in March 1979 when a G.E. sponsored Barbara Walters Special (an interview

*A Washington-based organization funded by businesses, trade groups and foundations to study media coverage of business and the economy.

with Jane Fonda and Tom Hayden and a clip from "The China Syndrome") was to air. G.E. dropped its sponsorship of the Special, it said, "because it contains material that could cause undue public concern about nuclear power."*

*The People's Almanac No. 2*** declares "control of the corporations that run the television networks and major sources of printed news is particularly important, because the news media influence the population as a whole. The top twenty-five newspapers that also own TV stations, book publishers, etc., have more than half of the daily newspaper circulation. Not only are most of the TV stations in the country associated with three media-conglomerate networks, but major banks hold controlling interests in each one. For instance, the trust departments of eleven banks hold a total of 38% of the stock in CBS; Chase Manhattan Bank, which holds 14% of CBS stock, also has substantial holdings in ABC and RCA, the parent of NBC." Those who control "media corporations do not dictate the news, but they place important constraints upon news managers and reporters."

The staff of the House Subcommittee on Domestic Finance wrote a report in 1968 on corporate interlocks in the U.S.

These typical pages from the two-volume, 1,945 page report (the size of a telephone book) show the board interlocks and/or stock ownership by First National City Bank in Westinghouse, G.E., twenty-one electric utilities, nine major publishing companies, ABC, Stone & Webster, and Allied Chemical, among others.

*Press Statement issued from G.E. corporate headquarters, Fairfield, Conn., dated February 23, 1979.

**A reference work by David Wallechinsky and Irving Wallace, William Morrow and Company, Inc., New York, 1978.

[SUBCOMMITTEE PRINT]

COMMERCIAL BANKS AND THEIR TRUST ACTIVITIES: EMERGING INFLUENCE ON THE AMERICAN ECONOMY

VOLUME 1

STAFF REPORT FOR THE SUBCOMMITTEE ON DOMESTIC FINANCE
COMMITTEE ON BANKING AND CURRENCY
HOUSE OF REPRESENTATIVES
90TH CONGRESS
2D SESSION

JULY 8, 1968

Printed for the use of the Subcommittee on Domestic Finance
Committee on Banking and Currency

U.S. GOVERNMENT PRINTING OFFICE
20-753 O WASHINGTON : 1968

TABLE 76.—INTERLOCKING RELATIONSHIPS BETWEEN FIRST NATIONAL CITY BANK, NEW YORK, N.Y., AND MAJOR CORPORATIONS

In Order By Standard Industrial Classification [1]

Classification by SIC code, and name of company	Director interlocks	Employee benefit funds managed by bank	Percent of outstanding stock held by bank [2]
Metal mining—nonproducers—SIC 107:			
Apex Minerals Corp			5. 3–C
Bituminous coal and lignite mining—SIC 121:			
Blue Diamond		1	15. 0–C
Crude petroleum and natural gas—SIC 131:			
Panoil Co		1	5. 2–C
General building contractors—SIC 151:			
Stone & Webster, Inc	1		
Canning and preserving fruits, vegetables—SIC 203:			
General Foods Corp	2	2	
Grain mill products—SIC 204:			
General Mills, Inc	1		
Alcoholic and malt beverages—SIC 208:			
National Distillers and Chemical Corp	1	2	12. 4–P 16. 4–P
Cigars—SIC 212:			
Consolidated Cigar Corp			6. 3–C
Textile mill products—SIC 221:			
Wyomissing Corp	1	2	
Paper & Allied Products—SIC 262:			
St. Regis Paper Co	2	1	
Kimberly-Clark Corp	1		
Boise Cascade Corp			18. 5–P
Potlatch Forests, Inc	1		
Building paper and building board mills—SIC 266:			
Upson Co	1		
Newspapers, periodicals, and books—SIC 271:			
McGraw-Hill, Inc			5. 8–P
Charles E. Merrill Books, Inc			12. 8–C
Prentice-Hall, Inc			8. 7–C
Wadsworth Publishing Co., Inc			10. 0–C
Allyn Bacon, Inc			7. 5–C
Harcourt, Brace & World, Inc			7. 6–C
American Book Co	1		
Meredith Publishing Co	1		
Doubleday & Co			23. 4–C
Industrial inorganic and organic chemicals—SIC 281:			
Monsanto Co	2	2	
W. R. Grace Co	3		
Allied Chemical Corp	1	1	
Celanese Corp			5. 6–P
Koppers Co., Inc	1		
Hooker Chemical Corp			6. 0–P

Footnotes at end of table, p. 716.

TABLE 76.—INTERLOCKING RELATIONSHIPS BETWEEN FIRST NATIONAL CITY BANK, NEW YORK, N.Y., AND MAJOR CORPORATIONS—Continued

In Order By Standard Industrial Classification [1]

Classification by SIC code and name of company	Director interlocks	Employee benefit funds managed by bank	Percent of outstanding stock held by bank [2]
Drugs—SIC 283:			
Bristol-Myers Co	1		
Upjohn Co			6. 1–C
Soap, detergents, and cleaning preparations—SIC 284:			
Procter & Gamble Co	1		
Colgate-Palmolive Co	1	2	
Petroleum refining—SIC 291:			
Standard Oil of New Jersey	1	1	
Mobil Oil Corp	2		
Phillips Petroleum Co		6	6. 6–C
Sinclair Oil Corp	2	1	
Glass and glass products—SIC 321:			
Owens-Illinois, Inc	1		
Corning Glass Works	2	4	8. 5–C
Concrete, gypsum, asbestos and plaster products—SIC 326:			
Johns-Manville Corp	2		
Blast furnances, steel works, and rolling and finishing mills—SIC 331:			
United States Steel Corp	1		
Dayton Malleable Iron Co			7. 8–C
Smelting and refining of nonferrous metals—SIC 333:			
Anaconda Co	2	2	
Reynolds Metals Co			7. 5–P
Kaiser Aluminum & Chemical Corp			7. 6–P
Kennecott Copper Corp	2		
Phelps Dodge Corp	1	1	
Scovill Manufacturing Co		1	15. 8–P
Arwood Corp	1		
Metal cans—SIC 341:			
American Can Co	2	1	
Farm machinery, construction, mining and materials handling machinery and equipment—SIC 352:			
Dresser Industries, Inc	1		
Metalworking machinery and equipment—SIC 354:			
Kearney & Trecker Corp			6. 5–C
Special industry machinery, excluding metalworking machinery—SIC 355:			
Ritter Pfaudler Corp	1		
Hobart Manufacturing Co			8. 0–C
General industrial machinery and equipment—SIC 356:			
Ingersoll-Rand Co	1		

Footnotes at end of table, p. 716.

TABLE 76.—INTERLOCKING RELATIONSHIPS BETWEEN FIRST NATIONAL CITY BANK, NEW YORK, N.Y., AND MAJOR CORPORATIONS—Continued

In Order By Standard Industrial Classification [1]

Classification by SIC code and name of company	Director interlocks	Employee benefit funds managed by bank	Percent of outstanding stock held by bank [2]
Office, computing, and accounting machines—SIC 357:			
International Business Machines Corp	1		
National Cash Register Co	2	1	
Service industry machines—SIC 358:			
Carrier Corp			11.5–P
Tecumseh Products Co			15.8–P
Electric transmission and distribution equipment—SIC 361:			
General Electric Co	1		
Westinghouse Electric Corp	1	2	6.6–P
Servel, Inc			7.5–P
Radio and television receiving sets—SIC 365:			
Magnavox Co	1		
Communication equipment—SIC 366:			
International Telephone & Telegraph	1	1	
Motor vehicles and motor vehicle equipment—SIC 371:			
Ford Motor Co	1		
Borg-Warner Corp	1	1	
Eaton, Yale & Towne, Inc			11.7–P
Mack Trucks, Inc			14.8–P
Aircraft and parts—SIC 372:			
Boeing Co	1	3	
United Aircraft Corp	2		
TRW, Inc			6.2–P
Railroad equipment—SIC 374:			
ACF Industries, Inc	1	1	
Optical instruments and lenses—SIC 383:			
Bell & Howell Co	1		
Xerox Corp	2		5.0–C
Toys, amusement, sporting goods, etc.—SIC 394:			
American Machine & Foundry Co			8.7–P
Jewelry, silverware, plated ware, etc.—SIC 391:			
Oneida, Ltd		1	5.5–C
Railroad transportation—SIC 401:			
Southern Pacific Co	1		
Union Pacific RR. Co	1		
Great Northern Ry. Co	1		
Northern Pacific Ry. Co	1		
Local and suburban passenger transportation—SIC 411:			
Trans-Caribbean Airways, Inc	1		
D.C. Transit System, Inc	1		

Footnotes at end of table, p. 716.

TABLE 76.—INTERLOCKING RELATIONSHIPS BETWEEN FIRST NATIONAL CITY BANK, NEW YORK, N.Y., AND MAJOR CORPORATIONS—Continued

In Order By Standard Industrial Classification [1]

Classification by SIC code and name of company	Director interlocks	Employee benefit funds managed by bank	Percent of outstanding stock held by bank [2]
Public warehousing—SIC 422:			
Merchants Refrigerating Co		3	10.2–C
Services incidental to water transportation—SIC 446:			
Coastal Ship Corp			11.2–C
Air transportation—SIC 451:			
United Air Lines, Inc			7.4–P
Pan American World Airways	1	2	
Telephone communication—SIC 481:			
American Telephone & Telegraph Corp	1		
Southern New England Telephone Co	1		
Commonwealth Telephone Co			19.1–P
New Jersey Bell Telephone Co	1		
New York Telephone Co	1		
Rochester Telephone Corp			10.0–P
			7.1–P
Hawaiian Telephone Co			5.0–P
			9.6–P
Wisconsin Telephone Co	1		
Ohio Bell Telephone Co	1		
Radio and TV broadcasting—SIC 483:			
ABC, Inc	1		
Communications services, not elsewhere classified—SIC 487:			
Communications Satellite Corp	1		
Electric companies and systems—SIC 491:			
Consolidated Edison Co. of New York, Inc	2	1	6.1–P
Southern California Edison Co			8.2–P
Virginia Electric & Power Co			5.0–P
Northern States Power			9.1–P
			5.8–P
			8.0–P
Long Island Lighting Co	1		8.2–P
Gulf States Utilities Co			7.2–P
Texas Power & Light Co			15.3–P
Connecticut Light & Power Co			5.8–P
Narragansett Electric Co			5.3–P
Ohio Power Co			11.3–P
Louisiana Power & Light Co			7.4–P
Dallas Power & Light Co			9.0–P
Texas Electric Service Co			5.3–P
Kansas City Power & Light Co			12.5–P
Florida Power Corp			5.1–P
			9.5–P
Arizona Public Service Co			11.8–P
Hawaiian Electric Co., Inc			13.1–P
			5.0–P
			25.3–P

Footnotes at end of table, p. 716.

TABLE 76.—INTERLOCKING RELATIONSHIPS BETWEEN FIRST NATIONAL CITY BANK, NEW YORK, N.Y., AND MAJOR CORPORATIONS—Continued

In Order By Standard Industrial Classification [1]

Classification by SIC code and name of company	Director interlocks	Employee benefit funds managed by bank	Percent of outstanding stock held by bank [2]
Gas companies and systems—SIC 492:			
Panhandle Eastern Pipe Line Co		4	9. 5–P
Southwest Gas Corp			8. 3–P
Intermountain Gas Co	1	1	
Washington Gas Light Co			6. 9–P
Northern Illinois Gas Co			7. 9–P
Tenneco, Inc			10. 9–P 11. 5–P 5. 8–P 6. 2–P
Colorado Interstate Gas Co			10. 3–P
Combination gas & electric systems—SIC 493:			
Public Service Electric & Gas Co			7. 3–P
Consumers Power Co	1	4	
Rochester Gas & Electric Corp	2		
Montana Dakota Utilities Co			6. 0–P
Department stores—SIC 531:			
Mercantile Stores Co., Inc	2	1	
Mail order houses—SIC 532:			
Sears, Roebuck & Co	1		
Vending machine operators—SIC 534:			
Canteen Corp	1		
Grocery and miscellaneous food stores—SIC 541:			
Food Fair Stores, Inc			6. 1–P
Jewel Companies, Inc		2	6. 0–C
Apparel and accessories stores, except shoes—SIC 561:			
J. C. Penney Co	2		
Shoe stores—SIC 566:			
Melville Shoe Corp			8. 0–P
Life, accident, and health insurance—SIC 631:			
Metropolitan Life	2		
New York Life	2		
Northwestern Mutual Life Insurance Co	1		
Travelers, Inc	1		
Mutual Life of New York	1		
United States Life Insurance Co	1		
Fire, marine, casualty, and surety insurance—SIC 633:			
Great American Insurance Co	1		
Federal Insurance Co	3		6. 7–C
Real estate—operators and lessors, except developers—SIC 651:			
City Investing Co	1	1	
General Real Estate Shares			7. 2–C

Footnotes at end of table, p. 716.

Of particular concern to the Congressional group were interlocks with media.

Said the report:

THE COMMERCIAL BANKS AND THE NEWS AND INFORMATION MEDIA

An area which should be of special concern because of its impact on public knowledge and opinion is the news and information media business. Several newspaper and magazine publishers have large blocks of stock held by commercial banks covered in the Subcommittee's survey. This includes 18 companies publishing 31 newspapers and 17 magazines, as well as operating 17 radio and TV stations.

Among the most prominent are Time, Inc., Newsday, Inc., The Evening News Association (Detroit), Booth Newspapers, Inc., The Tribune Company (Chicago), the Copley Press, Inc., The Hartford Courant Company, A. S. Abell Company (Baltimore Sun Papers), and The Dow Jones Company. The following table indicates the names of the publishing corporation, the newspapers or magazines published or radio and TV stations operated by it, and the name of the bank holding the stock, managing any of the company's pension funds or having a director interlock with the publishing corporation.

The percentages of common stock held and voted by some banks are indeed impressive and clearly constitute enough to control the corporation. For example, the Mercantile-Safe Deposit & Trust Co. of Baltimore, Maryland, holds 61.3 percent of the common stock of A. S. Abell Company, having sole voting rights over 27 percent of the total outstanding stock and partial voting rights over another 23.4 percent. In addition, Mercantile-Safe Deposit & Trust has three interlocking directorates with A. S. Abell Company. A. S. Abell is the owner and publisher of the *Baltimore Sun*, *The Evening Sun*, and the owner and operator of WMAR-TV, Baltimore, and WBOC-TV, Salisbury, Maryland. There are many other important examples shown in the following table.

TABLE 25.—INTERRELATIONSHIPS BETWEEN SURVEYED BANKS AND NEWS AND INFORMATION MEDIA

Publishers or operators of—	Name of bank	Director interlocks	Number of managed funds	Stockholding links and voting rights by percentage: Stock type [1]	Total percent of outstanding stock	Percent of sole voting right [2]	Percent of partial voting right [3]	Percent of no voting right [4]
New York Times Co.	Chase Manhattan Bank, N.A., New York, N.Y.	1						
	Morgan Guaranty Trust Co., New York, N.Y.		1	P	5.6		1.5	4.1
Total		1	1	P	5.6		1.5	4.1
Boston Herald Traveler Corp.	New England Merchant's National Bank, Boston Mass.	1						
	Old Colony Trust Co., Boston, Mass.	1	1					
Total		2	1					
Dow Jones Co., Inc.	Morgan Guaranty Trust Co., New York, N.Y.	1		C	9.7	8.0	.7	1.0
Booth Newpapers, Inc.	National Bank of Detroit, Mich.		1	C	9.6	3.9		5.7
	Detroit Bank & Trust, Detroit, Mich.			C	5.2	.8	4.4	
Total			1	C	14.8	4.7	4.4	5.7

Publishers or operators of—	Name of bank	Director interlocks	Number of managed funds	Stockholding links and voting rights by percentage: Stock type [1]	Total percent of outstanding stock	Percent of sole voting right [2]	Percent of partial voting right [3]	Percent of no voting right [4]
Time, Inc	First National Bank of Chicago, Ill.	1						
	Morgan Guaranty Trust Co., New York, N.Y.		1	C	8.1	4.1	7	3.3
	Chemical Bank New York Trust, New York, N.Y.	1						
Total		2	1	C	8.1	4.1	.7	3.3
McGraw Hill, Inc	New England Merchants National Bank, Boston, Mass.			P	5.5		1.2	4.3
	First National City Bank, New York, N.Y.			P	5.8	4.8	.1	.9
	Morgan Guaranty Trust Co., New York, N.Y.			P	9.9	1.1	8.8	
Meredith Publishing Co	First National City Bank, New York, N.Y.	1						
Prentice Hall, Inc	do			C	8.7	.8		7.9
Holt, Rinehart and Winston, Inc.	First National Bank of Chicago, Ill.			C	6.3	3.3	1.2	1.8
	Hartford National Bank & Trust, Hartford, Conn.	1						
Total		1		C	6.3	3.3	1.2	1.8
Ginn Co	First National Bank of Boston, Mass.	1						
Scott Foresman Co	State Street Bank & Trust, Boston, Mass.			C	5.7	0.1	0.2	5.4
	First National Bank of Chicago, Ill.	1	1	C	8.9	7.8	.7	.4
	Northern Trust Co., Chicago, Ill.			C	8.2	3.9	2.6	1.7
Total		1	1	C	22.8	11.8	3.5	7.5
Wiley, John, Sons, Inc	Morgan Guaranty Trust Co., New York, N.Y.	1	1	C	6.0	4.3		1.7
American Book Co	First National City Bank, New York, N.Y.	1						
Allyn Bacon, Inc	First National City Bank New York, N.Y.			C	7.5	.7		6.8
Penton Publishing Co	Central National Bank of Cleveland, Ohio.		4	C	6.3	3.0	3.3	
Simplicity Pattern Co., Inc.	Morgan Guaranty Trust Co., New York, N.Y.			P	15.8	9.9	1.0	4.9
	Bankers Trust Co., New York, N.Y.		1					
Total			1	P	15.8	9.9	1.0	4.9
Merrill, Charles E. Books, Inc.	First National City Bank New York, N.Y.			C	12.8	.3		12.5
Wadsworth Publishing Co., Inc.	First National City Bank New York, N.Y.			C	10.0			10.0
Butterick Co., Inc	Chemical Bank New York Trust, New York, N.Y.		1	C	39.0		23.4	15.6
Harcourt, Brace & World, Inc.	First National City Bank, New York, N.Y.			C	7.6	3.5	.6	3.5
	Morgan Guaranty Trust Co., New York, N.Y.	1		C	11.4	2.4	.4	8.6
Total		1		C	19.0	5.9	1.0	12.1

Publishers or operators of—	Name of bank	Director interlocks	Number of managed funds	Stockholding links and voting rights by percentage: Stock type [1]	Total percent of outstanding stock	Percent of sole voting right [2]	Percent of partial voting right [3]	Percent of no voting right [4]
Newsday, Inc.								
Newsday, Garden City, N.Y.	Morgan Guaranty Trust Co., New York, N.Y.			Class A, C.	49.0		49.0	
				Class B, C.	42.5		42.5	
				5% P	29.0		29.0	
Time, Inc.								
Time magazine, Life magazine, Fortune magazine, Sports Illustrated.	First National Bank of Chicago	1						
	Morgan Guaranty Trust Co., New York, N.Y.		1	C	8.1	4.1	0.7	3.3
	Chemical Bank N.Y. Trust, New York, N.Y.	1						
KLZ–TV—Denver, Colo.								
KLZ–AM–FM								
WOOD–TV—Grand Rapids, Mich.								
WOOD–AM–FM								
WFBM–TV—Indianapolis, Ind.								
WFBM–AM–FM								
KOGO–TV—San Diego, Calif.								
KOGO–AM–FM								
KERO–TV—Bakersfield, Calif.								
Holt, Rinehart, Winston, Inc.								
Field and Stream, Popular Gardening, New Homes Guide.	First National Bank of Chicago, Ill.			C	6.3	3.3	1.2	1.7
Home Modernizing Guide	Hartford National Bank & Trust Co., Hartford, Conn.	1						
Latrobe Printing & Publishing Co.								
Latrobe Bulletin, Latrobe, Pa.	Mellon National Bank & Trust Co., Pittsburgh, Pa.			C	48.2			48.2
Eagle Printing Co.								
Eagle, Butler, Pa.	Mellon National Bank & Trust Co., Pittsburgh, Pa.			Class A, C.	16.7		16.7	
				Class B, C.	13.0	(5)	(5)	
Daily News Publishing Co.								
News, McKeesport-Duquesne-Clinton, Pa.	Mellon National Bank & Trust Co., Pittsburgh, Pa.			C	33.4		33.4	
Booth Newspapers, Inc.								
News, Ann Arbor, Mich.	National Bank of Detroit, Detroit, Mich.		1	C	9.6	3.8		5.6
Times, Bay City, Mich.	Detroit Bank & Trust, Detroit, Mich.			C	5.2	.8	4.4	
Press, Grands Rapids, Mich.								
Journal, Flint, Mich.								
Citizen Patriot, Jackson, Mich.								
Chronicle, Muskegon, Mich.								
News, Saginaw, Mich.								

Publishers or operators of—	Name of bank	Director interlocks	Number of managed funds	Stockholding links and voting rights by percentage				
				Stock type [1]	Total percent of outstanding stock	Percent of sole voting right [2]	Percent of partial voting right [3]	Percent of no voting right [4]
D.A.C. News, Inc.								
D.A.C. News, Detroit, Mich.	National Bank of Detroit, Mich.			C	100.0		100.0	
The Tribune Co.								
Chicago Tribune, Chicago, Ill. WGN-TV—Chicago, Ill. Radio WGN KDAL-TV—Duluth, Minn. Radio KDAL	Continental Illinois National Bank, Chicago, Ill.			C	8.0	1.2	6.8	
Copley Press, Inc.								
Beacon News, Aurora, Ill.	First National Bank of Chicago, Ill.	1		C	100.0		100.0	
				Class B, C.	100.0		100.0	
Courier-News, Elgin, Ill. Herald-News, Joliet, Ill. State Journal, State Register, State Journal and Register, Springfield, Ill. Post Advocate, Alhambra, Calif. Union Tribune, San Diego, Calif.								
Hartford Courant Co.								
Hartford Courant, Hartford, Conn.	Connecticut Bank & Trust, Hartford, Conn.	2		C	92.0			92.0
	Hartford National Bank, Hartford, Conn.	2						
Evening News Association								
Detroit News, Detroit, Mich. WWJ-TV—Detroit, Mich. Radio WWJ-AM-FM	Detroit Bank & Trust, Detroit, Mich.			C	16.1	5.8		10.3
Chronicle Printing Co.								
Chronicle, Willimantic, Conn.	Connecticut Bank & Trust, Hartford, Conn.			C	66.0	66.0		
A. S. Abell Co.								
Baltimore Sun, Baltimore, Md.	Mercantile-Safe Deposit & Trust Co., Baltimore, Md.	3		C	61.3	27.0	23.4	11.9
Evening Sun, Baltimore, Md. WMAR-TV, Baltimore, Md. WBOC-TV, Salisbury, Md.	Maryland National Bank, Baltimore, Md.	1						

Publishers or operators of—	Name of bank	Director interlocks	Number of managed funds	Stockholding links and voting rights by percentage				
				Stock type [1]	Total percent of outstanding stock	Percent of sole voting right [2]	Percent of partial voting right [3]	Percent of no voting right [4]
Billboard Publishing Co.								
Billboard Magazine, Cincinnati, Ohio.	Central Trust Co., Cincinnati, Ohio.			C	59. 19	52. 61	6. 58	
				P	79. 50	65. 0	14. 5	
Western Hills Publishing Co.								
Western Hills Press, Western Hills, Ohio.	Central Trust Co., Cincinnati, Ohio.			C	64. 0	56. 6		7. 33
Dow Jones Co., Inc.								
Wall Street Journal, New York, N.Y. The National Observer, Silver Spring, Md. Barron's National Business and Financial Weekly, Chicopee, Mass.	Morgan Guaranty Trust Co., New York, N.Y.	1		C	9. 7	8. 0	0. 7	1. 0

The interlocks were found to have gone deep into the utility industry. Those identified include:

TABLE 23.—STOCKHOLDER, DIRECTORSHIP, AND EMPLOYEE BENEFIT FUND LINKS OF 49 BANKS SURVEYED WITH SERVICE INDUSTRY CORPORATIONS—Continued

Listed By Standard Industrial Classification*

Electric Companies and Systems, SIC Code 491

Name of company	Name of bank	Director interlocks	Employee benefit funds managed by bank	Voting arrangements				
				Stock type [1]	Total percent of outstanding stock	Percent sole vote [2]	Percent partial vote [3]	Percent no vote [4]
Consolidated Edison Co. of New York, Inc.	Chase Manhattan Bank, New York, N.Y.	1						
	First National City Bank, New York, N.Y.	2	1	P	6. 1			6. 1
	Morgan Guaranty Trust Co., New York, N.Y.	1	1	P	7. 0			
	Chemical Bank New York Trust, New York, N.Y.	**2**	**1**					
Total		**6**	**3**	**P**	**6. 1**			**6. 1**
				P	**7. 0**			
Commonwealth Edison Co.	**Continental Illinois National Bank, Chicago, Ill.**	**2**						
	First National Bank of Chicago, Illinois.	**3**						
	Northern Trust Co., Chicago, Ill.	**4**						
Total		**9**						

Name of company	Name of bank	Director interlocks	Employee benefit funds managed by bank	Voting arrangements				
				Stock type [1]	Total percent of outstanding stock	Percent sole vote [2]	Percent partial vote [3]	Percent no vote [4]
Southern California Edison Co.	National Shawmut Bank, Boston, Mass.			P	6.0			6.0
	First National City Bank, New York, N.Y.			P	8.2	0.3		7.9
Philadelphia Electric Co.	First Pennsylvania Bank & Trust, Philadelphia, Pa.	2						
	Philadelphia National Bank, Philadelphia, Pa.	2						
	Girard Trust Co., Philadelphia, Pa.	3		P	11.1	6.1	1.6	3.4
				P	5.4	3.1	1.7	.6
				C	5.9	2.1	2.2	1.6
				P	8.4	1.6	1.4	5.4
				P	9.1	5.1	3.3	.7
	Fidelity Bank, Philadelphia, Pa.	1		P	10.9	3.4		7.5
				P	12.2	10.9		1.3
				C	5.6	4.9	.1	.6
	Provident National Bank, Philadelphia, Pa.	1		C	5.6	1.2	2.2	2.0
Total		9		P	11.1	6.1	1.6	3.4
				P	5.4	3.1	1.7	.6
				C	17.1	8.2	4.5	4.2
				P	8.4	1.6	1.4	5.4
				P	9.1	5.1	3.3	.7
				P	10.9	3.4		7.5
				P	12.2	10.9		1.3
Niagara Mohawk Power Corp.	Morgan Guaranty Trust Co., New York, N.Y.	1	1					
Detroit Edison Co.	National Bank of Detroit, Mich.	2						
	Manufacturers National Bank of Detroit, Mich.	1						
	Detroit Bank & Trust, Detroit, Mich.	2						
Total		5						
Virginia Electric & Power Co.	First National City Bank, New York, N.Y.			P	5.0			5.0
				P	9.1	0.8		8.3
Union Electric Co.	First National Bank of Chicago, Ill.			P	5.0			5.0
Florida Power & Light Co.	Manufacturers Hanover Trust Co., New York, N.Y.	1						
	Morgan Guaranty Trust Co., New York, N.Y.		1	P	10.0			
				P	5.3			
Total		1	1	P	10.0			
				P	5.3			
Duke Power Co of North Carolina.	Morgan Guaranty Trust Co., New York, N.Y.	1	2					
Northern States Power	National Shawmut Bank, Boston, Mass.			P	7.3			7.3
	First National City Bank, New York, N.Y.			P	5.8			5.8
				P	8.0	.7	0.1	7.2
New England Electric System.	First National Bank of Boston, Mass.	1						
	Old Colony Trust Co., Boston, Mass.	2						
Total		3						
Long Island Lighting Co.	First National City Bank, New York, N.Y.	1		P	8.2			28.
	Morgan Guaranty Trust Co., New York, N.Y.			C	5.8	4.6	.5	.7
	Bankers Trust Co., New York, N.Y.	1						
Total		2		P	8.2			8.2
				C	5.8	4.6	.5	.7

Name of company	Name of bank	Director interlocks	Employee benefit funds managed by bank	Voting arrangements				
				Stock type [1]	Total percent of outstanding stock	Percent sole vote [2]	Percent partial vote [3]	Percent no vote [4]
Pacific Power & Light Co.	National Shawmut Bank, Boston, Mass.			P	6.6			6.6
Pennsylvania Power & Light Co.	Girard Trust Co., Philadelphia, Pa.			P	10.7	4.4	4.5	1.8
				P	14.4	9.2	5.1	.1
				P	5.7	2.9	2.2	.6
	Fidelity Bank, Philadelphia, Pa.			P	8.8	8.4		.4
	Mellon National Bank & Trust, Pittsburgh, Pa.	1	1					
Total		1	1	P	10.7	4.4	4.5	1.8
				P	14.4	9.2	5.1	.1
				P	5.7	2.9	2.2	.6
				P	8.8	8.4		.4
Potomac Electric Power Co.	National Shawmut Bank, Boston, Mass.			P	5.5			5.5
	First National Bank of Chicago, Ill.			P	5.3			5.3
Gulf States Utilities Co.	First National City Bank, New York, N.Y			P	7.2			7.2
	Morgan Guaranty Trust Co., New York, N.Y.	1						
Total		1		P	7.2			7.2
Illinois Power Co.	Continental Illinois National Bank, Chicago, Ill.	1						
	American National Bank & Trust, Chicago, Ill.	1						
Total		2						
Mississippi Power & Light Co.	National Shawmut Bank, Boston, Mass.			P	19.0			19.0
Georgia Power Co.	do			P	8.9			8.9
	First National Bank of Chicago, Ill.			P	5.3			5.3
	Girard Trust Co., Philadelphia, Pa.			P	5.0			5.0
				P	5.0			5.0
Texas Power & Light Co.	National Shawmut Bank, Boston, Mass.			P	11.3			11.3
				P	5.3			5.3
	First National City Bank, New York, N.Y.			P	15.3			15.3
Arkansas Power & Light Co.	National Shawmut Bank, Boston, Mass.			P	28.6			28.6
Otter Tail Power Co.	First National Bank of Chicago, Ill.			P	25.0			25.0
Connecticut Light & Power Co.	National Shawmut Bank, Boston, Mass.			P	11.3			11.3
	Connecticut Bank & Trust, Hartford, Conn.	1	1					
	First National City Bank, New York, N.Y.			P	5.8			5.8
Total		1	1	P	11.3			11.3
				P	5.8			5.8
United Illuminating Co.	First National Bank of Chicago, Ill.			P	9.3			9.3
Sierra Pacific Power Co.	do			P	15.0			15.0
Hartford Electric Light Co.	National Shawmut Bank, Boston, Mass.			P	13.5			13.5
				P	16.3			16.3
	First National Bank of Chicago, Ill.			P	8.7			8.7
	Hartford National Bank & Trust, Hartford, Conn.	3						
	Connecticut Bank & Trust, Hartford, Conn.	1						
Total		4		P	13.5			13.5
				P	16.3			16.3
				P	8.7			8.7

Name of company	Name of bank	Director interlocks	Employee benefit funds managed by bank	Stock type [1]	Voting arrangements: Total percent of outstanding stock	Per cent sole vote [2]	Per cent partial vote [3]	Per cent no vote [4]
Holyoke Water Power Co.	Connecticut Bank & Trust, Hartford, Conn.	1		C	8.0	1.7		6.3
	Bankers Trust Co., New York, N.Y.	1						
Total		2		C	8.0	1.7		6.3
Public Service Co. of New Mexico.	First National Bank of Chicago, Ill.			C	5.3	4.1	0.2	1.0
				P	7.7			7.7
Narragansett Electric Co.	First National Bank of Boston, Mass.	1						
	First National City Bank, New York, N.Y.			P	5.3			5.3
Total		1		P	5.3			5.3
Public Service Co. of New Hampshire.	Old Colony Trust Co., Boston, Mass.	1						
Ohio Power Co.	First National City Bank, New York, N.Y.			P	11.3			11.3
Eastern Utilities Associates.	Old Colony Trust Co., Boston, Mass.	1						
New England Power Co.	First National Bank of Boston, Mass.	1						
	First National Bank of Chicago, Ill.			P	5.5			5.5
Total		1		P	5.5			5.5
Louisiana Power & Light Co.	First National City Bank, New York, N.Y.			P	7.4			7.4
	Morgan Guaranty Trust Co., New York, N.Y.			P	7.1			
	Girard Trust Co., Philadelphia, Pa.			P	5.0			5.0
Central Vermont Public Service Corp.	Old Colony Trust Co., Boston, Mass.	1						
Dallas Power & Light Co.	National Shawmut Bank, Boston, Mass.			P	7.0			7.0
	First National City Bank, New York, N.Y.			P	9.0			9.0
Central Louisiana Electric Co., Inc.	First National Bank of Chicago, Ill.			P	10.7			10.7
	Morgan Guaranty Trust Co., New York, N.Y.			C	6.0	5.9		.1
Upper Peninsula Power Co.	Cleveland Trust Co., Cleveland, Ohio.	1						
Texas Electric Service Co.	National Shawmut Bank, Boston, Mass.			P	15.0			15.0
	First National City Bank, New York, N.Y.			P	5.3			5.3
	Morgan Guaranty Trust Co., New York, N.Y.			P	6.3			6.3
West Penn Power Co.	Girard Trust Co., Philadelphia, Pa.			P	6.1	3.2	2.6	.3
				P	17.1	9.7	2.1	5.3
				P	17.2	12.8	3.9	.5
Metropolitan Edison Co.	do			P	10.9	4.8	4.4	1.7
				P	10.2	3.0	4.2	3.0
				P	17.0	10.8	5.1	1.1
	Fidelity Bank, Philadelphia, Pa.			P	6.8	6.7		.1
				P	9.2	8.1		1.1
Kansas City Power & Light Co.	Harris Trust & Savings Bank, Chicago, Ill.			P	5.7			
	First National City Bank, New York, N.Y.			P	12.5	.4	.1	12.0
	Morgan Guaranty Trust Co., New York, N.Y.			P	5.0			
				P	6.5			6.5

Name of company	Name of bank	Director interlocks	Employee benefit funds managed by bank	Stock type [1]	Voting arrangements: Total percent of outstanding stock	Percent sole vote [2]	Percent partial vote [3]	Percent no vote [4]
Duquesne Light Co	Girard Trust Co., Philadelphia, Pa.			P	11.3	8.1	2.8	.4
				P	7.9	7.1	.8	
	Mellon National Bank & Trust, Pittsburgh, Pa.	2	1					
Total		2	1	P	11.3	8.1	2.8	.4
				P	7.9	7.1	.8	
Oklahoma Gas & Electric Co.	Morgan Guaranty Trust Co., New York, N.Y.			P	13.1			
Philadelphia Electric Power Co.	First Pennsylvania Bank & Trust, Philadelphia, Pa.	1						
	Philadelphia National Bank, Philadelphia, Pa.	1						
	Provident National Bank, Philadelphia, Pa.	1						
Total		3						
Boston Edison Co	First National Bank of Boston, Mass.	3						
	State Street Bank & Trust, Boston, Mass.	1						
	National Shawmut Bank, Boston, Mass.	1						
	Old Colony Trust Co., Boston, Mass.	2	2					
Total		7	2					
Cleveland Electric Illuminating Co.	Cleveland Trust Co., Cleveland, Ohio.	2	1					
	National City Bank of Cleveland, Ohio.	2						
	Central National Bank of Cleveland, Ohio.	1						
	Society National Bank of Cleveland, Ohio.	1						
Total		6	1					
New York State Electric & Gas Corp.	State Street Bank & Trust, Boston, Mass.			C	5.3	.2		5.1
	Morgan Guaranty Trust Co., New York, N.Y.			P	11.9			
Southwestern Electric Power Co.	First National Bank of Chicago, Ill.			P	10.0			10.0
Alabama Power Co	National Shawmut Bank, Boston, Mass.			P	12.4	0.4		12.0
				P	14.2			14.2
				P	6.2			6.2
	Girard Trust Co., Philadelphia, Pa.			P	7.1			7.1
Idaho Power Co	Old Colony Trust Co., Boston, Mass.			C	6.1	2.8	1.1	2.2
	Mellon National Bank & Trust, Pittsburgh, Pa.		1	C	5.5	4.7	.2	.6
Total			1	C	11.6	7.5	1.3	2.8
New Jersey Power & Light Co.	Girard Trust Co., Philadelphia, Pa.			P	10.8	8.4	1.9	.5
Central Hudson Gas & Electric.	Old Colony Trust Co., Boston, Mass.			C	28.9	2.0		26.9
	Mellon National Bank & Trust, Pittsburgh, Pa.			P	8.3			8.3
Florida Power Corp	National Shawmut Bank, Boston, Mass.			P	6.1			6.1
	First National Bank of Chicago, Ill.			P	5.0			5.0
	First National City Bank, New York, N.Y.			P	5.1			5.1
				P	9.5			9.5
	Morgan Guaranty Trust Co., New York, N.Y.			P	5.8			
	Girard Trust Co., Philadelphia, Pa.			P	10.2	.2		10.0

Name of company	Name of bank	Director interlocks	Employee benefit funds managed by bank	Voting arrangements: Stock type [1]	Total percent of outstanding stock	Percent sole vote [2]	Percent partial vote [3]	Percent no vote [4]
Pennsylvania Electric Co.	**First National Bank of Chicago, Ill.**			P	14.7			14.7
	Morgan Guaranty Trust Co., New York, N.Y.			P	10.0			
	Girard Trust Co., Philadelphia, Pa.			P	13.4	5.3	4.3	3.8
				P	7.0	5.3	1.5	.2
				P	7.5	1.3	.4	5.8
				P	18.2	9.6	3.5	5.1
	Fidelity Bank, Philadelphia, Pa.			P	11.1	10.6		.5
Atlantic City Electric Co.	**National Shawmut Bank, Boston, Mass.**			P	6.0			6.0
Missouri Public Service Co.	State Street Bank & Trust, Boston, Mass.	1						
	National Shawmut Bank, Boston, Mass.			P	10.0			10.0
Total		1		P	10.0			10.0
Utah Power & Light Co.	National Shawmut Bank, Boston, Mass.			P	5.2			5.2
				P	5.5			5.5
				P	17.0			17.0
	First National Bank of Chicago, Ill.			P	15.2			15.2
Indiana & Michigan Electric Co.	do			P	14.2			14.2
Interstate Power Co.	National Shawmut Bank, Boston, Mass.			P	21.8			21.8
Southwestern Public Service Co.	State Street Bank & Trust, Boston, Mass.	1						
	Philadelphia National Bank, Philadelphia, Pa.	1						
Total		2						
Arizona Public Service Co.	National Shawmut Bank, Boston, Mass.			P	5.0			5.0
	First National Bank of Chicago, Ill.			P	10.0			10.0
	First National City Bank, New York, N.Y.			P	11.8			11.8
Hawaiian Electric Co., Inc.	First National Bank of Chicago, Ill.			P	20.0			20.0
				P	14.3			14.3
	First National City Bank, New York, N.Y.			P	13.1			13.1
				P	5.0			5.0
				P	25.3			25.3
Tampa Electric Co.	National Shawmut Bank, Boston, Mass.			P	11.0			11.0
	New England Merchants National, Boston, Mass.	1						
	First National Bank of Chicago, Ill.			P	10.0			10.0
Total		1		P	11.0			11.0
				P	10.0			10.0
Savannah Electric & Power Co.	National Shawmut Bank, Boston, Mass.			P	10.2			10.2
Central Maine Power Co.	Old Colony Trust Co., Boston, Mass.	1						
Monongahela Power Co., Ohio.	Mellon National Bank, Pittsburgh, Pa.	4						
Northeast Utilities	Hartford National Bank & Trust, Hartford, Conn.	2	1					
	Connecticut Bank & Trust, Hartford, Conn.	1						
	United Bank & Trust, Hartford, Conn.	1						
Total		4	1					

Declared Wright Patman, the Congressional committee's chairman:

LETTER OF TRANSMITTAL

JULY 8, 1968.

This study reveals that a major shift in the control of American business has been developing over the last 25 years. The implications of this fundamental shift in the structure of our economy are very far-reaching. Further study will be required to fully comprehend them. But definitely this shift has implications for the continued viability of our competitive free enterprise economy. It raises questions of conflict of interest for individuals and bank trustees; of anticompetitive activity for banks, other financial institutions and corporations; and of basic problems of public policy which touch on some of the most fundamental political and economic issues of our time for the Congress, the Executive and the Judiciary.

While further examination and discussion of these major issues will undoubtedly be called for before this Subcommittee and elsewhere, it is my view that the data presented here for the first time show that the American economy of today is in the greatest danger of being dominated by a handful of corporations in a single industry as it has been since the great money trusts of the early 1900's. Through the various devices described in this study, commercial banks control the investments of billions of dollars of funds and vote large blocks of stock of major corporations in practically every important industry in the economy. These same banking institutions have gained representation on boards of directors of and serve as major sources of credit for many of these same major industrial corporations. Therefore, a few banking institutions are in a position to exercise significant influence, and perhaps even control, over some of the largest business enterprises in the nation. In addition, thousands of small and medium-sized businesses all over the country are dominated by the major commercial banks in their areas. The pervasiveness of the banks' position cannot be denied by any reasonable person who studies the data in this report. The question that the Congress and the American people must decide is whether this situation should be allowed to continue and develop further without substantial legislative and administrative checks to this concentration of economic power in the hands of a few.

Perhaps the words of President Wilson, uttered over 50 years ago, should stand today as a warning to the American public as they did at that time:

> The great monopoly in this country is the money monopoly. So long as that exists, our old variety and freedom and individual energy of development are out of the question. A great industrial nation is controlled by its system of credit. Our system of credit is concentrated. The growth of the nation, therefore, and all our activities are in the hands of a few men who, even if their actions be honest and intended for the public interest, are necessarily concentrated upon the great undertakings in which their own money is involved and who, necessarily, by every reason of their own limitations, chill and check and destroy genuine economic freedom. This is the greatest question of all, and to this statesmen must address themselves with an earnest determination to serve the long future and the true liberties of men.

Wright Patman

WRIGHT PATMAN, *Chairman.*

Concluded the study:

COMMERICIAL BANKS AND THEIR TRUST ACTIVITIES: EMERGING INFLUENCE ON THE AMERICAN ECONOMY

SUMMARY AND CONCLUSIONS

The problem of concentration of economic wealth has been of concern to Congress and the public for many years. On occasion this concern has led to the enactment of important legislation, particularly between 1890 and 1914, and again during the New Deal Era.

Since World War II the nature of the problem has changed considerably, in large part because of the dramatic growth of institutionally managed funds, including mutual funds, insurance funds, employee benefit funds, and private trust funds held by bank trust departments. It appears that the trend identified in the 1930's of major corporations in the United States being controlled by corporate management because of the wide dispersal of stock ownership among large segments of the public, may now be giving way to a new trend toward control of these vital elements of our economy through control of the voting of large blocks of stock in these corporations held for beneficiaries by a relatively few giant financial institutions.

In addition to the general discussion of the activities of the trust departments of the 49 banks surveyed, and the interrelationships between these banks and other corporations on an industry basis, the Subcommittee survey makes possible a detailed examination of these activities on a city-by-city basis for the 10 cities surveyed. These 49 banks reported that they had a total of 8,019 director interlocks with 6,591 companies, an average of 164 director interlocks with an average of 135 companies per bank. These 49 banks also reported the names of 5,270 companies in which they hold individually 5 percent or more of the outstanding shares of one or more classes of stock, an average of 108 companies per bank.

This clearly puts each of these banks in a position of great influence over a large segment of business in their areas. More important, it gives the banking community enormous potential power, for good or evil, over important parts of the nation's corporate structure.

There are many factors to be considered in judging the potential for influence and control that banks and other financial institutions may have over other corporations. These include the supply of capital, the holding and voting of large blocks of stock of companies and extensive interlocking directorships between financial institutions and other corporations. All of these factors, as well as many others, appear to exist to an extensive degree at the present time. One of the problems of detecting this situation in the case of commercial banks has been their favored position as contrasted with other institutional investors in being able to conceal the extent to which they hold and control investments of other corporations.

Some of the consequences of bank influence over major corporations involve potential restraint of competition, both among financial institutions, and between competing nonfinancial corporations which may be linked together through a single banking institution. Mergers, which would not otherwise take place, may also be fostered by the influential position of banks with one or more companies involved. The ties between banks and other corporations may also unduly restrict the sources of credit available to competing businesses which do not have the same links with banks. This, too, is a form of restraint on competition.

In addition, there are a number of serious conflict of interest problems that arise from extensive interrelationships between banks and other corporations.

What this all adds up to is that the major banking institutions in this country are emerging as the single most important force in the economy, both through the huge overall financial resources at their command and through the concentration of these resources and other interrelationships with a large part of the nonbanking business community in the country. Earlier reports have discussed both the trend toward concentration within commercial banking itself during the post-war period and—even more significantly—the growing interlocking relationships between these major banking institutions and other major financial institutions, such as insurance companies and mutual savings banks. The power of the banks alone is quite impressive. In combination with these other financial institutions it would be overwhelming.

When the power of these financial institutions, in the combination which appears to be evolving, is examined in connection with their power—both existing and potential—over a large part of the nonfinancial sectors of our economy, the picture is complete. The kind of snowballing economic power described in this study, with its literally thousands of interlocking relationships, is a situation which can only be ignored at great peril.

Though it is "a situation which can only be ignored at great peril," as the Congressional report declares, the growing interlocking pattern of the American economic structure has largely been ignored. And since the report was issued things have become even more interlocked and centralized.

In the 1970's there is the possibility that getting interlocked under the structure was government itself. In 1973, David Rockefeller and a chief foreign policy advisor, Zbigniew Brzezinski, formed what was to be called The Trilateral Commission.

It is what investigative reporter Craig Karpel describes as "the closest thing possible to a board of directors of the world." Rockefeller and Brzezinski put together an "action committee" made up of the top executives of the most important multinational corporations and financial institutions ranging from J. Paul Austin, chairman of The Coca-Cola Company and Peter G. Peterson, chairman of Lehman Brothers, to political figures: from Rockefeller family advisor Henry Kissinger to Democrat Jimmy Carter and Representative John Anderson and, at one point, Republican George Bush; and including representatives of labor, education and media.

This is a recent listing of North American members of The Trilateral Commission as it appears on its literature:

The Trilateral Commission

GEORGES BERTHOIN *European Chairman*	TAKESHI WATANABE *Japanese Chairman*	DAVID ROCKEFELLER *North American Chairman*
EGIDIO ORTONA *European Deputy Chairman*		MITCHELL SHARP *North American Deputy Chairman*
	GEORGE S. FRANKLIN *Coordinator*	
HANNS W. MAULL *European Secretary*	TADASHI YAMAMOTO *Japanese Secretary*	CHARLES B. HECK *North American Secretary*

North American Members

*I. W. Abel, *Former President, United Steelworkers of America*

David M. Abshire, *Chairman, Georgetown University Center for Strategic and International Studies*

Gardner Ackley, *Henry Carter Adams University Professor of Political Economy, University of Michigan*

Graham Allison, *Dean, John F. Kennedy School of Government, Harvard University*
Doris Anderson, *Former Editor,* Châtelaine *Magazine*
John B. Anderson. *House of Representatives*
Anne Armstrong, *Former U.S. Ambassador to Great Britain*
J. Paul Austin, *Chairman, The Coca-Cola Company*
George W. Ball, *Senior Partner, Lehman Brothers*
Michel Belanger, *President, Provincial Bank of Canada*
*Robert W. Bonner, Q.C., *Chairman, British Columbia Hydro*
John Brademas, *House of Representatives*
Andrew Brimmer, *President, Brimmer & Company, Inc.*
William E. Brock, III, *Chairman, Republican National Committee*
Arthur F. Burns, *Senior Adviser, Lazard Frères & Co.; former Chairman of Board of Governors, U.S. Federal Reserve Board*
Hugh Calkins, *Partner, Jones, Day, Reavis & Pogue*
Claude Castonguay, *President, Fonds Laurentien; Chairman of the Board, Imperial Life Assurance Company; former Minister in the Quebec Government*
Sol Chaikin, *President, International Ladies Garment Workers Union*
William S. Cohen, *House of Representatives*
*William T. Coleman, Jr., *Senior Partner, O'Melveny & Myers; former Secretary of Transportation*
Barber B. Conable, Jr., *House of Representatives*
John Cowles, Jr., *Chairman, Minneapolis Star & Tribune Co.*
Alan Cranston, *United States Senate*
John C. Culver, *United States Senate*
Gerald L. Curtis, *Director, East Asian Institute, Columbia University*
Lloyd N. Cutler, *Partner, Wilmer, Cutler & Pickering*
Louis A. Desrochers, *Partner, McCuaig and Desrochers, Edmonton*
Peter Dobell, *Director, Parliamentary Centre for Foreign Affairs and Foreign Trade, Ottawa*
Hedley Donovan, *Editor-in-Chief, Time Inc.*
Claude A. Edwards, *Member, Public Service Staff Relations Board; former President, Public Service Alliance of Canada*
Daniel J. Evans, *President, The Evergreen State College; former Governor of Washington*
Gordon Fairweather, *Chief Commissioner, Canadian Human Rights Commission*
Thomas S. Foley, *House of Representatives*
George S. Franklin, *Coordinator, The Trilateral Commission; former Executive Director, Council on Foreign Relations*
Donald M. Fraser, *House of Representatives*
John Allen Fraser, *Member of Parliament, Ottawa*
John H. Glenn, Jr., *United States Senate*
Donald Southam Harvie, *Deputy Chairman, Petro Canada*
Philip M. Hawley, *President, Carter Hawley Hale Stores, Inc.*
Walter W. Heller, *Regents' Professor of Economics, University of Minnesota*
William A. Hewitt, *Chairman, Deere & Company*
Carla A. Hills, *Senior Resident Partner, Latham, Watkins & Hills; former U.S. Secretary of Housing and Urban Development*
Alan Hockin, *Executive Vice President, Toronto-Dominion Bank*
James F. Hoge, Jr., *Chief Editor,* Chicago Sun Times
Hendrik S. Houthakker, *Henry Lee Professor of Economics, Harvard University*

Thomas L. Hughes, *President, Carnegie Endowment for International Peace*
*Robert S. Ingersoll, *Deputy Chairman of the Board of Trustees, The University of Chicago; former Deputy Secretary of State*
D. Gale Johnson, *Provost, The University of Chicago*
Edgar F. Kaiser, Jr., *President and Chief Executive Officer, Kaiser Resources Ltd.*
Michael Kirby, *President, Institute for Research on Public Policy, Montreal*
Lane Kirkland, *Secretary-Treasurer, AFL-CIO*
*Henry A. Kissinger, *Former Secretary of State*
Sol M. Linowitz, *Senior Partner, Coudert Brothers; former Ambassador to the Organization of American States*
Winston Lord, *President, Council on Foreign Relations*
Donald S. Macdonald, *Former Canadian Minister of Finance*
*Bruce K. MacLaury, *President, The Brookings Institution*
Paul W. McCracken, *Edmund Ezra Day Professor of Business Administration, University of Michigan*
Arjay Miller, *Dean, Graduate School of Business, Stanford University*
Lee L. Morgan, *President, Caterpillar Tractor Company*
Kenneth D. Naden, *President, National Council of Farmer Cooperatives*
David Packard, *Chairman, Hewlett-Packard Company*
Gerald L. Parsky, *Partner, Gibson, Dunn & Crutcher; former Assistant Secretary of the Treasury for International Affairs*
William R. Pearce, *Vice President, Cargill Incorporated*
Peter G. Peterson, *Chairman, Lehman Brothers*
Edwin O. Reischauer, *University Professor and Director of Japan Institute, Harvard University; former U.S. Ambassador to Japan*
*Charles W. Robinson, *Vice Chairman, Blyth Eastman Dillon & Co.; former Deputy Secretary of State*
*David Rockefeller, *Chairman, The Chase Manhattan Bank, N.A.*
John D. Rockefeller, IV, *Governor of West Virginia*
Robert V. Roosa, *Partner, Brown Bros., Harriman & Company*
*William M. Roth, *Roth Properties*
William V. Roth, Jr., *United States Senate*
John C. Sawhill, *President, New York University; former Administrator, Federal Energy Administration*
Henry B. Schacht, *Chairman, Cummins Engine Inc.*
*William W. Scranton, *Former Governor of Pennsylvania; former U.S. Ambassador to the United Nations*
*Mitchell Sharp, *Member of Parliament; former Minister of External Affairs*
Mark Shepherd, Jr., *Chairman, Texas Instruments Incorporated*
Edson W. Spencer, *President and Chief Executive Officer, Honeywell Inc.*
Robert Taft, Jr., *Partner, Taft, Stettinius & Hollister*
Arthur R. Taylor
James R. Thompson, *Governor of Illinois*
Russell E. Train, *Former Administrator, U.S. Environmental Protection Agency*
Philip H. Trezise, *Former Assistant Secretary of State for Economic Affairs*
Paul A. Volcker, *President, Federal Reserve Bank of New York*
Martha R. Wallace, *Executive Director, The Henry Luce Foundation, Inc.*
Martin J. Ward, *President, United Association of Journeymen and Apprentices of the Plumbing and Pipe Fitting Industry of the United States and Canada*
Glenn E. Watts, *President, Communications Workers of America*
Caspar W. Weinberger, *Vice President and General Counsel, Bechtel Corporation*
George Weyerhaeuser, *President and Chief Executive Officer, Weyerhaeuser Company*

Marina v.N. Whitman, *Distinguished Public Service Professor of Economics, University of Pittsburgh*
Carroll L. Wilson, *Mitsui Professor in Problems of Contemporary Technology, Alfred P. Sloan School of Management; Director, Workshop on Alternative Energy Strategies, MIT*
T. A. Wilson, *Chairman of the Board, The Boeing Company*

*Executive Committee

Former Members in Public Service

Lucy Wilson Benson, *U.S. Under Secretary of State for Security Assistance*
W. Michael Blumenthal, *U.S. Secretary of the Treasury*
Robert R. Bowie, *U.S. Deputy to the Director of Central Intelligence for National Intelligence*
Harold Brown, *U.S. Secretary of Defense*
Zbigniew Brzezinski, *U.S. Assistant to the President for National Security Affairs*
Jimmy Carter, *President of the United States*
Warren Christopher, *U.S. Deputy Secretary of State*
Richard N. Cooper, *U.S. Under Secretary of State for Economic Affairs*
Richard N. Gardner, *U.S. Ambassador to Italy*
Richard Holbrooke, *U.S. Assistant Secretary of State for East Asian and Pacific Affairs*
Walter F. Mondale, *Vice President of the United States*
Henry Owen, *Special Representative of the President for Economic Summits; U.S. Ambassador at Large*
Jean-Luc Pépin, P.C., *Cochairman, Task Force on Canadian Unity*
Elliot L. Richardson, *U.S. Ambassador at Large with Responsibility for UN Law of the Sea Conference*
Gerard C. Smith, *U.S. Ambassador at Large for Non-Proliferation Matters*
Anthony M. Solomon, *U.S. Under Secretary of the Treasury for Monetary Affairs*
Cyrus R. Vance, *U.S. Secretary of State*
Paul C. Warnke, *Director, U.S. Arms Control and Disarmament Agency; Chief Disarmament Negotiator*
Andrew Young, *U.S. Ambassador to the United Nations*

The last series of names, "Former Members in Public Service," is, as you will notice, about all of the top members of the Carter administration–virtually every major post in the Departments of State, Defense and the Treasury and, of course, the President himself and his Vice President.

Jimmy Carter had become an original member of The Trilateral Commission after coming to the Rockefeller family's attention while he was Governor of Georgia. David Rockefeller, chairman of Chase Manhattan Bank, has multi-million dollar personal investments around Atlanta, Ga., sometimes called "Rockefeller Center South."

"The key to Rockefeller and Brzezinski's plan was to gain control of the executive branch of the U.S. government. The first step was the selection in 1973 of an ambitious, capable presidential candidate," wrote Karpel.*

With the Republican Party plagued by Watergate and seen as a sure loser in the 1976 presidential election, "David [Rockefeller] and Zbig [Brzezinski] had both agreed that Carter was the ideal politician to build on," Carter's former deputy campaign chief, Peter Bourne, wrote.**

"The Democratic candidate will have to emphasize work, the family, religion, and increasingly, patriotism, if he has any desire to be elected," said Brzezinski in 1973 as director of the Trilateral Commission.

Carter gave his support to Rockefeller and Brzezinski's notion of a system of unrestricted multinational trade. Brzezinski's major work, *Between Two Ages*, describing such trade and his concept of a global "technetronic" society is what attracted the Rockefellers to him. Importantly, Carter was an enthusiast of the energy form that both the Rockefellers and Brzezinski saw fueling the multinational future they sought: nuclear power.

A former member of Admiral Hyman Rickover's Division of Naval Reactors, Carter has described his profession as "an engineer and nuclear physicist."

With Brzezinski writing all of Carter's foreign policy speeches, according to Karpel, and with support by other Trilateral members and from the media conglomerates managed by Trilateral members, what next happened was–as Karpel writes–"the presidency of the United States and the key cabinet departments of the federal government" were "taken over by a private organization dedicated to the subordination of the domestic interests of the United States to the international interests of the multinational banks and corporations. This seizure of public power by private interests is the most serious political scandal in American history. Watergate was someone named Martinez breaking into the office of the Democratic National Committee in the dead of night. Cartergate is David Rockefeller breaking into the Oval Office in broad daylight."

Carter, Karpel stresses, often does the opposite of what he promises. In his acceptance speech at the Democratic National Convention in 1976 he denounced "unholy, self-perpetuating alliances [that] have

*In a series of articles published in *Penthouse Magazine.* Reprints are available from the magazine at 909 Third Avenue, New York, N.Y. 10022.
***Far Eastern Economic Review,* May 1977.

been formed between money and politics . . . a political and economic elite who have shaped decisions and never had to account for mistakes nor to suffer from injustice. When unemployment prevails, they never stand in line looking for a job. When deprivation results from a confused . . . welfare system, they never go without food or a place to sleep. When the public schools are inferior or torn by strife, their children to go exclusive private schools. And when the bureaucracy is bloated and confused, the powerful always manage to discover and occupy niches of special influence and privilege." As Karpel writes, "No one has ever given a more eloquent description of the membership of The Trilateral Commission."

He would declare in the first Presidential candidates' debate in 1976 that we should "use atomic energy only as a last resort, with the strictest possible safety precautions." But once in office it was "nuclear power must play an important role in the U.S. to insure our energy future."

Indeed, Carter seems to be following the prescription of another Trilateral Commission report* which declares, "Once he is elected president, the president's electoral coalition has . . . served its purpose. The day after his election the size of his majority is almost . . . entirely irrelevant to his ability to govern . . . . The governing coalition need have little relation to the electoral coalition."

Karpel's thesis may be correct; the policy laid out by The Trilateral Commission appears to be the policy of the Carter administration. This can be seen by anyone going through the policy declarations of The Trilateral Commission.

The Trilateral program is laid out in a series of reports called "The Triangle Papers."** On the energy issue, "The Triangle Papers: 17, *Energy: Managing the Transition,"* describes the road we're on.

*"The Triangle Paper. 8, *The Crisis of Democracy."* 1975.

**The group will send copies of their publications for a fee. Write to its headquarters at 345 East 46th Street, New York, N.Y. 10017

The Trilateral Commission

345 EAST 46th STREET, NEW YORK, N.Y. 10017 • (212) 661-1180
Cable: TRILACOM NEWYORK • Telex: 424787

October 20, 1978

MEMORANDUM

SUBJECT: Energy: Managing the Transition

FROM: George Franklin, Coordinator GF
Charles Heck, North American Secretary CBH

Enclosed is the published report of the Trilateral Energy Task Force. While the report remains the responsibility of its three authors -- Messrs. John Sawhill, Hanns Maull and Keichi Oshima -- its preparation involved extensive international consultations with energy advisers and experts from the trilateral areas as well as from the OPEC and nonoil-producing developing countries.

You will note that the report emphasizes the need for a coordinated trilateral management to assure the orderliness of the long-term transition to a new generation of energy technologies. More specifically, its recommends 1) that each trilateral government raise domestic energy prices to world market levels (and perhaps beyond in some cases); 2) increased cooperation between the OECD and OPEC countries; 3) provision of financial and technical assistance to less developed countries to develop their indigenous energy resources; and 4) the development of a concerted trilateral policy on nuclear energy and nuclear weapons proliferation as rapidly as possible.

A six-page summary of the report is included on pages ix-xiv. We hope this report will prove of interest.

EUROPEAN OFFICE: 151, boulevard Haussmann, 75008 Paris, France
JAPANESE OFFICE: Japan Center for International Exchange, 4-9-17 Minami-Azabu, Minato-ku, Tokyo, Japan

△ THE TRIANGLE PAPERS: 17

ENERGY:

MANAGING THE TRANSITION

John C. Sawhill
President
New York University

Keichi Oshima
Professor
of Nuclear Engineering
University of Tokyo

Hanns W. Maull
European Secretary
The Trilateral Commission

The Trilateral Commission

A Private North American-European-Japanese
Initiative on Matters of Common Concern

The report, co-authored by John Sawhill, a Trilateral Commission member, board member of Con Edison and currently Deputy Undersecretary of Energy, starts off with what is described as the "oil shock" of 1973—apparently we are undergoing oil shock treatments—and lays down domestic and foreign energy policy. It places great faith in nuclear power. Says the report:

> Nuclear plants are large, highly capital-intensive energy sources primarily useful for providing base load generating capacity, and in some countries, they will undoubtedly play an increasingly important role in this sector of the electric utility industry. Yet, citizens groups in almost every one of the Trilateral countries have raised serious questions about their government's nuclear program, claiming that the current generation of fission reactors creates certain undesirable environmental effects and has the added disadvantage of producing a highly-radioactive waste that must be stored for thousands of years. Some of these protests have been quite vocal and have had the effect of limiting the number of sites available for future plants, increasing the time required to build new plants, and, as a consequence, increasing the cost and reducing the potential economic benefits of nuclear energy.
>
> This "social constraint" is a relatively new dimension in the nuclear equation that governments must take into account in evaluating their future nuclear plans. The existence of such a constraint will make it increasingly necessary for governments to encourage open discussion and debate on nuclear issues and to disseminate as widely as possible correct information on the benefits and safety of nuclear power in comparison with the risks associated with it.

In late March of 1979 came the Three Mile Island accident, and in Tokyo in April 1979 (shortly before Carter and leaders of industrial nations were to meet there) The Trilateral Commission gathered to hear from the architects of its energy program. According to *Trialogue,* the Commission's quarterly magazine:

THE ENERGY REPORT REVISITED

• *Delays in Nuclear Programs*

A great number of participants noted that the Three Mile Island nuclear accident has inevitably raised very serious questions on the future of nuclear energy—in the United States itself, but also in most other trilateral countries. Consequently, it will be more difficult for these countries to pursue vigorous nuclear programs. Some North American participants expressed their alarm at what they viewed as a "meltdown" of the credibility of our experts in the wake of the accident. Others suggested that this accident should be seen as a useful warning and as a positive development, since neither a break in the reactor shield nor a big radioactivity release occurred.

On the other hand, several Europeans and Japanese, noting the delays which are likely to affect their countries' efforts to develop nuclear power, emphasized that Japan and Europe have "no other option" than moving ahead with their nuclear programs. Stressing the need for a common strategy in this area on the part of the trilateral countries, and the political courage which it will take for our leaders to shift the debate from the safety issue to the energy supply issue, some of these participants insisted that our countries "stop interfering with each other's policies" in the nuclear area, and urged that a strong pronouncement to this effect be made at the June 1979 Tokyo Summit meeting.

Notice "the political courage which it will take for our leaders to shift the debate from the safety issue to the energy supply issue" after Three Mile Island. "Our leaders" did shift the debate: within weeks we felt the 1979 "oil shock," a deliberately created oil shortage.

It "may well go down in history as one of the greatest frauds ever perpetrated on a helpless people." declared Fred J. Cook in *The Nation*.* He documented his report with numerous government reports which confirmed that there was no shortage of oil. It was a " 'crisis' carefully orchestrated by Big Oil, aided and abetted by the complacent non-watchdog in the Department of Energy. And even by President Carter himself," Cook wrote.

The president of Holiday Inns, Inc. was explaining why he wasn't concerned for his motel chain because the fuel shortage was but a "government induced media scare designed to get support for a broad-based energy program."

There was great public resentment, the "crisis" eased, and the oil companies could hardly hold in their profits (Exxon surpassed General Motors as the world's major industrial company). Now as the 1980's begin, shortages are again to be expected–perhaps black-outs and brown-outs and gas rationing–to continue an energy crisis atmosphere, as in the old Standard Oil industrial strike technique, forcing up prices, knocking out independents and, importantly, serving to promote nuclear power. The AEC in an early booklet on "the energy crisis," framed it this way:

> The energy crisis is serious but it can be met—and at a price that is financially and environmentally feasible.

> We will need to meet our increasing demands for electricity primarily by nuclear reactors particularly the breeder reactor. For the more distant future beyond the year 2000, it will be necessary to develop nuclear fusion, solar power, and other exotic systems for producing and storing energy.

*July 28, 1979.

If you would like more detailed information on this subject, please send for copies of the information booklets "Nuclear Power Plants", "Electricity and Man", "Atomic Energy and Your World", and "Breeder Reactors".

And while talking "energy crisis," the government declares:

1. Central-station atomic power development is being pursued as a matter of both national and industrial policy;

2. Radiation protection regulations have been laid down by competent authority to govern this activity; *

Notice, nuclear power is being *pursued as a matter of both national and industrial policy.*

*From "Atomic Power Safety," U.S. Atomic Energy Commission, Washington, D.C., 1964.

Of great concern to The Trilateral Commission is freedom of the press. Declares *The Crisis of Democracy:* "The responsibility of the press should now be increased to be commensurate with its power; significant measures are required to restore an appropriate balance between the press, the government, and other institutions in society."

Indeed, Brzezinski in *Between Two Ages* proposes a "world information grid, for which Japan, Western Europe, and the United States are most suited," a grid that "could create the basis for a common educational program, for the adoption of common academic standards, for the organized pooling of information, and for a more rational division of labor in research and development."

The Crisis of Democracy, issued in 1975, calls for "prior restraint" of what the media may publish in unspecified "unusual circumstances" and for government to have "the right and ability to withold the information at the source" and its "moving promptly to reinstate the law of libel as a necessary and appropriate check upon the abuses of power by the press."

So here we are with a strong nuclear industrial establishment firmly in control and strong indications that the Standard Oil empire is centrally involved, and using its traditional tactics at full power—with a "nuclear engineer" up front, as President of the United States.

And things can be expected to get rougher. As Dan Sheehan, attorney in the Karen Silkwood case, in which the revelations concerning Ms. Silkwood's death led to a jury awarding $10.5 million to her survivors, said at *The Village Voice* 1979 "teach-in" on nuclear power:

"The folks that are telling you that nuclear power is safe and that there's only going to be perhaps one cancer death from TMI are the same ones who told the American public that they had things under control just before the Tet offensive. These are the same people who kept constantly being able to perceive the alleged light at the end of that everlasting tunnel in Vietnam. It is important to understand that we are not dealing with a good faith effort to communicate the truth to us. Now that might sound a bit facetious, but I believe personally that there's the major obstacle that we have to overcome. It takes a different type of consciousness to know that the people in the executive department of the United States government are intentionally, willfully, and I'll use the word, conspiratorially, deceiving the American public."

The "same group" that former President Dwight Eisenhower warned about in his farewell speech is involved, the same group "we were up against when we tried to stop the Vietnam war," said Sheehan. "There is an industrial complex that has seized the control of the executive department of the U.S. government and it is run by the big businesses and the corporations of America and they will not tolerate this information getting out."

"In the Karen Silkwood case, we have proved beyond the shadow of any reasonable doubt," continued Sheehan, "that Karen Silkwood was in physical possession of X-rays that were required to be taken of those fuel rods that were being brought to the Hanford facility (from the Kerr-McGee plutonium plant) for the fast-flux test breeder reactor. She could prove that the corporation out there in Oklahoma was intentionally covering over flaws in those welds of those fuel rods with magic markers. She had possession of those documents and she was on her way to bring them to David Burnham when she was killed. Those documents were taken from that car by agents of the Kerr-McGee Corporation and they were in the presence of AEC officials at the time they took the documents. Now, don't you believe for one single instant that this is not going to be a war on the domestic front as serious as the Vietnam War . . . . There is a deceitful, intentional, willful campaign underway to keep this industry in place. There is no rational reason for it being there other than pure profit. And that's the thing we have to deal with. . . . In the Karen Silkwood case we have now proved that Karen Silkwood had electronic eavesdropping equipment placed in her home, she had wiretaps on her telephones and that the American government officials in the executive department were up to here in that particular conspiracy."

"The private industrial complex that is utilizing this nuclear power is knowingly and willfully taking plutonium and special nuclear materials from those facilities with the knowledge of the executive department of the United States and bringing those bomb grade materials to such countries as, it used to be Iran, and it is South Africa and it is Brazil. Special nuclear materials are being taken out of those facilities for the purpose of arming those countries with nuclear weapons. That's a fact, and the U.S. Congress knows it. The CIA knows it. . . . Does anybody here know that? No. Does anybody here think that all we have to do is ask them to stop that and they'll stop it? Absolutely not."

"The fact of the matter is we have come to the knowledge that we are being deceived on the body counts, we're being deceived on the costs, we're being deceived on the basic reasons why this thing continues."

"There is at this moment in place in the United States a full-scale clandestine surveillance system funded by the U.S. executive department to spy on the people who are attempting to bring this industry to a halt," Sheehan went on. "Now don't expect the Justice Department to bring any indictment. The Justice Department is not our Justice Department, the Justice Department is the Justice Department which is there to protect those people who run those industries for that profit. It's going to take a different state of mind on the part of the American people to bring this thing to a halt."

According to the evidence developed by Sheehan in the Silkwood case, "the people who did the surveillance" of Ms. Silkwood were "trained and equipped" at a Florida operation that receives federal funds and is run by former CIA agents. It is the same place, he says, where Michael Townley and Cuban refugees who assassinated the Chilean ambassador to the U.S. were trained. "This is one of the most important links that we've hit into in the Karen Silkwood case because now we know the type of people that were involved are the type of people who are perfectly capable of knocking people off." The federal government is working through such operations, says Sheehan, "because under the federal civil rights act, as a rule, we can get at the government agencies in discovery and sue these people. . . . The reason the federal government started to transfer some things into private agencies is because they think they're immune against lawsuit for private conspiracies against civil rights."

He speaks of a similar operation in Georgia also providing "special training to undertake illegal surveillance," also led by a former CIA operative–a former member of "an assassination team that was trained by the CIA to go into Cuba." This operation features seven "specially equipped automobiles that you could turn a little switch on their dashboard and it will change the configuration of their headlights for following people at night." The group had a list of nuclear plant workers regarded as "potential dissidents," adds Sheehan, because they had "questioned safety procedures." Private information about these individuals was being researched. "It's not just Karen Silkwood," concludes Sheehan. "It's going on in Georgia, it's going on in Florida," indeed all over the U.S., with spying operations in California and in New England undertaking "surveillance" of the Clamshell Alliance there.

Further, Sheehan speaks of a "highly secret organization" within the federal government the Silkwood case has "uncovered," called the Defense Industrial Security Command "which works with industries in the United States that it believes are 'necessary to the national defense.' "

A police state structure is seen as the direction of a nuclear-powered society, to prevent "proliferation"–just a few pounds of plutonium from any reactor or spent storage pool or fabricating plant and a terrorist or hoodlum or madman has enough to make an atomic bomb.

But the Silkwood case points to police state measures being used in conjunction with the continued great push for a nuclear-powered society.

How did we get so far? As Harvard Business School teacher Irvin Bupp explains: "The theology of nuclear power and the sanctification of light water technology created an interlocking set of intellectual, political and commercial interests. Scientists with an intellectual stake in the success of nuclear power, politicians with a political stake, bureaucrats with an organizational stake, and businessmen with a commercial stake reinforced and amplified each other's claims. Much of this misinformation appears to have its roots in the early American mistake of fitting nuclear power development into the client-patron pattern of government. By serving as soapboxes for the economic claims of the reactor manufacturers, the Atomic Energy Commission and the Joint Committee amplified the flow of misinformation and decisively altered the strength of these companies at home and abroad."

Bupp and Jean-Claude Derian wrote a book, *Light Water, How The Nuclear Dream Dissolved,* about how America and its nuclear program spread through the world.

"Turnkey" contracts offered by U.S. reactor manufacturers and a belief in "American technical superiority" were main factors, they say. American reactor manufacturers' claims were "accepted" as "general faith in American technological omnipotence." They note: "In retrospect many Europeans might be inclined to be bitter since they have apparently been the victims of the same confusion between expectation and fact that originated on the other side of the Atlantic Ocean."

And eighty per cent of reactors in the world are now American made, principally by Westinghouse and General Electric.

In the U.S. sometimes it is said, "We have to keep up with the world on nuclear power." In fact, the world has been hooked by the U.S. on nuclear power. No nation has nearly as many nuclear power plants planned as the U.S.

Here is a chart of current and planned nuclear plants.

Countries with Nuclear Power Reactors by 1984

Country	Number of power reactors	Country	Number of power reactors
Argentina	2	Korea	3
Austria	1	Mexico	2
Belgium	8	Netherlands	2
Brazil	3	Pakistan	1
Bulgaria	4	Philippines	1
Canada	19	Poland	1
China	?	Romania	1
Czechoslovakia	5	S. Africa	2
Finland	4	Spain	18
France	38	Sweden	12
FRG	36	Switzerland	9
GDR	5	Taiwan	6
Hungary	2	UK	39
India	7	USA	165
Iran	4	USSR	35
Italy	9	Yugoslavia	1
Japan	32		

Source: *Nuclear Engineering International*, April 1977, Supplement.

Note: Some nuclear power plants consist of two (or more) reactor units—hence, for example, the 39 reactors predicted for the UK in this table.

Although the Standard Oil/Rockefeller/Trilateral configuration appears to be foremost within it, the nuclear establishment is a wider complex. Nuclear power as a concept arrived at the same time as did a tendency toward centralization of power in industrial society in general–from electric power to political power. In the U.S. it is not only Carter but Ronald Reagan, John Anderson, George Bush, Gerald Ford, and virtually the entire power structure who support nuclear power. Similarly, in the Soviet Union, the power structure strenuously moves that nation into reliance on nuclear energy. So go France and Germany. Despite differences in political ideology, means of industrial production are copied–and America, as a result of an extraordinary sequence of events, has set the pattern.*

Further, the catalyzing forces of nuclear scientists, engineers and technicians have become international, functioning "in the way of a religious cult," explains Lorna Salzman of Friends of the Earth. They would comprise a "nuclear priesthood with the secrets to this technology and we must believe them and follow them and, in exchange for promises of material benefits, we will give up our individual political power and our souls." It is the "cult of nuclear technology." And it has been promulgated widely.

"Men can do stupid things," Alexander Cockburn and James Ridgeway have written. "From the sixth to the fourth centuries B.C., the leaders of the expanding Athenian empire disposed of their natural resources in a way so disastrous that the consequences are still being felt today along the shores of the Mediterranean. Military imperatives required the construction of ever larger number of ships. Across the centuries, the forests of Attica, Cyprus, and Sicily slowly receded in deference to these imperial requirements. By the fourth century, Plato himself was complaining that the forests of ancient memory were gone and that the scrub that remained could scarcely support the humble honey bee. By the time the woodman's axe was stilled, the frantic Greeks–their lands eroded and barren–were offering tax credits to those farmers attempting to replenish the land. It was too

*The League of Conservation Voters of Washington, D C. gives Carter a C+ on nuclear power and gives nearly all the other 1980 major presidential contenders marks going down from there, based on their records on the issue.

Ronald Reagan got the lowest rating–an F. Reagan, who has declared "we shouldn't worry too much about the near-meltdown of Three Mile Island," insists nuclear power is "essential" and that the country has "no choice" but to go to nuclear power. "The emotional campaign against nuclear power plants not only exaggerates the hazards of any such power to generate electricity but is equally irrational in its advocacy of solar power as a substitute," he said in a 1979 radio spot. For many years he was what General Electric describes as the company's "general good will ambassador," touring the nation for the firm as well as serving as the host on G.E. Theatre on television. "When I go on tour for the company I make as many as 14 talks a day to various groups," said Reagan in 1960, adding: "One pretty high government official tried to have me shut up once, but the president of G E. told me to go ahead and say whatever I wanted to say."

John Anderson, the Republican-turned-Independent candidate for president in 1980, is described by the League as "one of the leading proponents of nuclear power in the House of Representatives. For 14 years he was a member of the Congressional Joint Committee on Atomic Energy, the League notes, and he has declared "we must forge a link between the future of nuclear power and the safety of nuclear power." He is a Trilateral Commission member.

And George Bush, a nuclear booster also with ties to the Trilateral Commission, received a D on nuclear power from the League of Conservation Voters.

Only Senator Edward Kennedy of the major party hopefuls received a good grade from the League on nuclear power–an A-.

late. The land remains barren today. Two thousand years from now, the mutant inheritors of the earth–assuming that there are any–will peer through all three eyes at the consequences of human folly in the 20th century; abandoned power plants still radioactive for 80,000 years; poisoned water, poisoned lands. How, they will ask themselves, could 20th-century people, amply endowed with renewable resources and rich stocks of fossil fuels, have been so stupid as to place their trust in nuclear energy?"*

**Village Voice,* May 7, 1979.

CHAPTER EIGHT

The Alternatives

Beyond everything else, nuclear power is not needed. We have an array of safer, cheaper, more sensible alternatives.

The way the nuclear industrial establishment carries on, you'd think we're dependent on nuclear power. In fact, only three per cent of the total end use of energy comes from nuclear power. Nuclear power can only be used to generate electricity, and electricity is one of the smallest components of the energy picture–it provides but thirteen per cent of total energy use.

If all the nuclear plants in America were closed down right now, it wouldn't make a bit of difference. Electric systems have reserve capacity–in the U.S. thirty-five to forty-five per cent excess reserve. Taking away the thirteen per cent nuclear power contributes to the American electric system would still leave the system with a twenty-three per cent-plus excess capacity.

Nuclear power has been pushed with the justification that even though it has risks, we need the energy it provides. Not only don't we need nuclear-electric energy now, but its maximum possible contribution could never be a principal energy supply.

Still, because of the vast power of the nuclear industrial establishment, our energy approach has been thrown completely out of kilter. We are and have been spending billions upon billions of dollars–enriching the energy corporations and nuclear bureaucrats and technocrats–for what isn't truly needed. And as Denis Hayes, director of the U.S. Solar Energy Research Institute, declares: "The capital nature of nuclear development will foreclose other options." If we continue to dump the bulk of our energy capital into an energy source which at best can't return that investment in proportion, we'll really be left with an energy crisis.

"For some applications," explains energy planner and physicist Amory Lovins, "electricity is appropriate and indispensible: electronics, smelting, subways, most lighting, some kinds of mechanical work, and a few more. But these uses are already oversupplied."

The U.S. "end uses that really require electricity" could be "reduced to five per cent, and these could be handled with the present U.S. hydroelectric capacity plus co-generation"–the generation by industry of on-site electricity with the heat it otherwise wastes. "Thus an affluent industrial economy," says Lovins, "could advantageously operate with no central power stations at all."

"People do not want electricity or oil, nor such economic abstractions as 'residential services', but rather comfortable rooms, light, vehicular motion, food, tables and other real things," explains Lovins. "In the United States today, about 58 per cent of all energy at the point of end use is required as heat, split roughly 25-35 between temperatures above and below the boiling point of water. In Western Europe the low temperature heat alone is often half of all end-use energy. Another 38 per cent of all U.S. end use energy provides mechanical motion: 31 per cent in vehicles, 3 per cent in pipelines, 4 per cent in industrial electric motors. The rest, a mere 4 per cent of delivered energy, represents all lighting, electronics, telecommunications, electrometallurgy, electrochemistry, arc welding, electric motors in home appliances and in railways, and similar end uses that now require electricity.

"In short," says Lovins, "our energy supply problem is overwhelmingly–about 90 per cent–a problem of heat and portable liquid fuels" and these "needs can be met much more cheaply, quickly and easily without going through electricity."

As to the claim that nuclear power is needed as a substitute for "expensive imported oil," this is a farce which the nuclear industry can be expected to continue to promote by applying pressure at the gasoline pumps to try to brainwash people into thinking there's a connection. But no more than sixteen per cent of electricity in America or Europe is generated by oil–it is produced primarily with coal, and in certain places with hydropower, gas and geothermal energy. (And the oil used to generate electricity is residual oil, the densest, heaviest of oil types which can't be used in cars or in heating oil. Its main use is in power plants.)

Further, points out Lovins, the electricity "market is so small" that substituting nuclear power for the little electricity produced by oil "can make little difference to the oil problem whether one is concerned with the next ten years or the next hundred. Nuclear power cannot leap the boundaries of this narrow market–cannot become more than a 4 per cent term in total end-use energy." He concludes:

"Therefore nuclear power is not necessary and indeed is a positive encumbrance whose resource, political and infrastructural commitments get in the way of what we should be doing instead. If nuclear power is unnecessary and uneconomic then we need not argue about whether it is safe and otherwise acceptable . . . . These issues become irrelevant. All that remains is to devise an orderly terminal phase for an unfortunate aberration."

We are now at a crossroads in energy decision-making.

In one direction is an appropriate blend of safe, here-now energy forms–a combination of solar power, wind power, geothermal energy, power from waste and co-generation, power from plants, among many other sources–all of these plus energy efficiency.

It is what Lovins calls the "soft energy path." By soft he doesn't mean "vague, mushy, speculative, or ephemeral, but rather flexible, resilient, sustainable, and benign."

Solar power alone is capable of satisfying much of the major end use of energy: heat. Some 1.5 quadrillion megawatt-hours of solar energy arrive at the earth's outer atmosphere each year–an amount 28,000 times greater than all the commercial energy used by humankind. The sun, which in twelve hours sends down to the United States the nation's yearly consumption of energy, is available as a permanent substitute for oil (after the world's brief bout with petroleum for heat) and for much more.

"About one-fifth of all energy used around the world now comes from solar resources: wind power, water power, biomass, and direct sunlight" explains Denis Hayes, a top U.S. government figure on solar power. "By the year 2000, such renewable energy sources could provide 40 per cent of the global energy budget; by 2025, humanity could obtain 75 per cent of its energy from solar resources."

"This timetable would require an unprecedented worldwide commitment of resources and talent," says Hayes, but "every essential feature of the proposed solar transition has already proven technically viable" and "if the fifty-year timetable is not met, the roadblocks will have been political–not technical."

Much of the power for locomotion, mechanical motion, can come from liquid and gaseous fuels derived from biomass sources–as Brazil is now demonstrating, converting its vehicular sector to run on homegrown alcohol rather than gasoline.

And there are many more energy sources on the soft energy path–ways through which we can have all the energy we need and not get killed in the process.

The other energy road is one Lovins associates with "hard energy." It is one with an emphasis on nuclear power and "costly, complex, centralized and gigantic plants," wasteful, environmentally damaging, lethally dangerous–and having no connection with real needs.

This is the road we've been shoved onto–for the sake of the short term profits of the nuclear establishment.

The U.S. Senate Committee on Small Business conducted an important investigation in 1975 into why the United States was neglecting the development of solar power. It found that the government and the nuclear industry had joined so that large corporations could maximize profits through nuclear power.

"The suspicion is almost unavoidable," said Gaylord Nelson, U.S. senator from Wisconsin and committee chairman, "that the absurdly low estimates of the solar contributions during the next 25 years are projections not of what the estimators think this country could do . . . but rather what they hope is the most the country will do . . . . Not because doing so little is in the best interests of the majority of Americans and other people of the world, but because doing more could possibly threaten existing investments in other technologies. The giant firms, which have large investments in nuclear power, hope that solar energy will not gain rapidly."

"Nuclear technology is big-business technology," says Raymond D. Watts who was general counsel to the Senate committee. "Solar technology, however, is uniquely suited to small business."

U.S. Senator Thomas J. McIntyre who co-chaired the committee declared: "I believe that this enslavement–coupled with a blind refusal to research and develop clean, renewable alternative energy sources–must, indeed, be one of the most calamitous errors in human judgement since time began."

The report of the committee noted:

> the committee questioned why for years solar energy had been virtually ignored as an alternative energy resource—even in the face of promising Government reports dating back to the early 1950's. In 1952, the Paley Commission, in its final report to President Truman, stated: "Efforts made to date to harness solar energy economically are infinitesimal. It is time for aggressive research in the whole field of solar energy—an effort in which the United States could make an immense contribution to the welfare of the free world." [3]
>
> The amount of energy from the sun reaching the United States annually, according to the Paley Commission, was 1,500 times the nation's energy consumption in 1950.[4]

But, the Senate panel went on,

Tragically, the Paley Commission's recommendations "for aggressive research" went unheeded. As a result, the Commission's optimistic prediction for solar engery's contribution to the national energy budget was unrealized.

The committee maintained:

Anything that slows down the development of solar energy—the one cheap, limitless source of energy that cannot be shut down by war or embargo—is undermining the national security.

After its series of hearings, the Senate panel declared: "Within the near future, solar power could and certainly should contribute significantly to the national energy budget."

It said:

Had the United States Government followed the recommendations of the Paley Commission in 1952, "for aggressive research in the whole field of solar energy," it seems probable that the country might now be achieving the solar equivalent of—and thereby saving for other uses or for future generations—3 million barrels of oil (or oil equivalent) per day, or more.[43] While the fossil fuels we could thus have saved but didn't are now burned forever, the opportunity for future savings is even greater, and the need more compelling, now than in 1952. The nation could and should establish immediately the goal of providing at least 30 percent of its building heating and cooling, and water heating from the sun by 2000, with significant percentage increases each year from now until then.

The commercialization of a non-polluting, environmentally acceptable and cost-effective energy system is not a challenge for the distant future. The challenge is now. The United States has a growing solar heating and cooling industry which can markedly accelerate its manufacturing capability provided the proper incentives are there. In fact, a massive research and development program may be unnecessary, if not counterproductive, to the extent that it could further postpone the wide-scale use of solar heating and cooling systems.

Until very recently, the Federal government has not accepted that challenge. By ignoring the recommendations of the Paley Commission, it left solar energy up to the small business pioneers and individual innovators who, acting on their own initiative, with virtually no government support, were responsible for almost all of the solar energy research, development, and demonstration work that occurred in this country prior to 1973.

Now that the Federal government has decided to embark on an accelerated program for solar energy development, one might think that the pioneers would finally get their rightful share of participation, but that has rarely been the case. The Federal departments and agencies charged with the development of solar energy have not adequately considered the needs and capabilities of small business. The agencies have not sufficiently consulted the Small Business Administration, have not established small business set-asides, and have usually relied on and favored big business concerns and giant universities.

The committee's report spoke of

overt or covert opposition to rapid solar development by various institutions, ranging from the oil industry and the electric utilities to the Federal government.

and it recommended that the government

should concentrate more of its energies and funds on smaller, decentralized applications of solar energy, many of which are already proven, less expensive to implement, less prone to large-scale blackouts or other failures, and less likely to lead to the establishment of anti-competitive and concentrated conditions in the emerging solar energy industries.

Receiving major government contracts to estimate the potential of solar energy, the committee learned, were General Electric and Westinghouse. It noted that General Electric in a 1974 report claimed that only 1.6 per cent of America's energy supply could come from solar energy by the year 2000* and Westinghouse said 3.04 per cent.**

Meanwhile, just the year before, several contrary reports had been made by the government's own National Science Foundation. One NSF 1073 study, "Solar Energy Program Report," declared, "Ultimately, practical solar energy systems could easily contribute 15 to 30 per cent of the nation's energy requirements." Another report,

*General Electric Co., "Solar Heating and Cooling of Buildings: Phase O, Feasibility and Planning Study, Final Report," May 1974.

**Westinghouse Electric Corporation, "Solar Heating and Cooling of Buildings: Phase O, Final Report," May 1974.

"Solar Energy Research Program Alternatives," projected a thirty-five per cent contribution by the year 2000. Yet another, "Solar Energy as a National Energy Resource," declared, "solar energy can be developed to meet sizable portions of the nation's future energy needs."

Much of what little money has been put into solar development by the government—one/three thousandths of the amount it spent for nuclear development in 1970, a typical year—has also gone to Westinghouse and General Electric as well as Mobil, Union Carbide and other major corporate members of the nuclear industry.

Another tactic the industry has employed: nuclear firms, for many years, have been busy buying up budding solar energy companies.

"The fossil/nuclear industry is fighting the rapid development" of solar energy because of the "serious threat" solar energy poses to it, testified Edwin Rothschild of Consumers Solar Electric Power Corporation before the Federal Energy Administration in 1975. "This explains, in part, why Exxon, Mobil and Shell have bought out small solar electric power companies. This explains, in part, the government's long time-frame for the development on a commercial scale of solar electric power. This explains, in part, why the larger, more powerful companies, especially those which have the most to lose, are getting many of the federal grants in solar energy research and development." He concluded: "I do not, however, believe that the government can continue to cover up the sun."

John Keyes has written a book entitled *The Solar Conspiracy** on the attempt at corporate domination and suppression of solar energy. The move is no different from attempts through the centuries of powerful people and organizations to control resources, but he concludes: "Those who would try to implement the control of the sun should beware, because this time the game went too far. They've toyed with that principle they should not have—stealing that which every man knows in his instinctual heart to be his own . . . the sun."

Scott Denman and Ken Bossong, staff associates at the Citizens Energy Project, in 1979 noted: "In a vicious cycle, the money continues to be donated to the same companies that want to hold back solar as long as possible and then make it as expensive—and bring it under the same monopoly control—as our current energy forms. So far, what little money the government spends on solar research is being handed to corporate giants on a silver platter." **

*Morgan & Morgan, Dobbs Ferry, N.Y., 1975.
***Seven Days,* June 19, 1979.

They point to Honeywell, "a major research firm with an interlocking directorate of executives from the nuclear, oil, natural gas and banking industries" having been the "big winner in 1977" to get the government contract "to develop a Transportable Solar Laboratory (TSL) which would travel around the country, explaining the techniques and virtues of solar energy at public workshops. Instead Honeywell used the TSL to discourage solar use, especially the low-cost methods. Furious, the California State Energy Commission 'strongly recommended' that the tour be 'immediately stopped and never started again', because of 'misleading analysis techniques . . . and a manifest lack of knowledge'. But the TSL continued, still under Honeywell's auspices."

E.F. Schumacher has said: "With the rise in the importance of solar energy, we have the rare opportunity of either standing by and watching an attempt to create a new monopoly before our eyes, or we can add our support in an effort to see that solar energy is developed and used for our best benefits as individuals, as a society and as a world."

John Berger, former energy projects director of Friends of the Earth, puts it his way: "A number of corporations with large nuclear commitments have belittled solar power in an effort to buy time for the redemption of their unwise nuclear investments" and "federal energy policy has furthered their ends." Berger stresses, "We have no crisis in energy supply . . . . We have enough energy available from the sun, the wind, the oceans, and the nation's large coal reserves to provide far more energy than we need–now or in the foreseeable future. We live in a giant and virtually inexhaustible energy flux that is more than adequate for our needs. Yet we do have a real crisis–a crisis in energy policy, and a closely related environmental crisis . . . . The crisis is evident in the government's failure to invest adequate resources in either clean energy systems or in energy conservation . . . . The U.S. energy bureaucracy is unresponsive to citizens' needs for safe, monopoly-free energy because it is overly responsive to the pervasive pressures of giant energy corporations that wield such dangerous influence on our government. Our current energy policy is an expression of the large energy corporations' desire not to upset the current economic-political status quo, which rests on rapid energy growth rates and extravagant energy use."*

Integral to the soft energy path is energy efficiency.

For starters, half the energy used in the United States is wasted,

**Nuclear Power: The Unviable Option,* Ramparts Press, Palo Alto, California, 1976.

and this does not contribute to better living a bit. Sweden, Denmark, and Switzerland consume only one-half the per-capita energy of the U.S. and each of these nations has a higher per-capita gross national product than America. Energy efficiency "means doing better, not doing without," says Ralph Nader.

"The notion that energy efficiency and renewable energy sources will entail 'hardship' and 'sacrifice' applies only to utilities and energy corporations which thrive on profit from energy waste, overconsumption and construction of unnecessary energy facilities," declares Lorna Salzman.

As the Harvard Business School said in its 1979 report, *Energy Future,* "There is a source of energy that produces no radioactive waste, nothing in the way of petrodollars, and very little pollution," and energy efficiency "is no less an energy alternative than oil, gas, coal or nuclear. Indeed, in the near term, conservation could do more than any of the conventional sources . . . . If the United States were to make a serious commitment to conservation, it might well consume 30 to 40 per cent less energy than it now does and still enjoy the same or an even higher standard of living." Energy efficiency was described by the Harvard report as "the key energy source."

"Dollar for dollar, investments in increasing the energy efficiency of buildings, industries and the transportation system will save more energy than expenditures on new energy facilities will produce," says Hayes.

Energy efficiency is using the right type and scale of energy for the work to be done, and minimizing waste. For instance, central electric plants waste two-thirds of the energy they generate—and the one-third that is left often goes to heat houses electrically. "Where we want only to create temperature differences of tens of degrees, we should meet the need with sources whose potential is tens or hundreds of degrees, not with a flame temperature of thousands or a nuclear equivalent to trillions—like cutting butter with a chainsaw," says Lovins.

The Harvard Business School projection that America could use thirty to forty per cent less energy and that this would not make one iota of difference to the quality of life is widely confirmed. "A 30 to 40 per cent reduction in energy use is entirely feasible," says Dr. George Kistiakowsky, formerly President Kennedy's science advisor. Dr. Robert H. Williams, senior scientist for the Ford Energy Project and Dr. Marc H. Ross, director of the Center for Environmental Studies at Princeton University, say forty-five per cent of U.S. energy could be saved. And the Ford Energy Project said in its 1974 report,

A Time to Choose: "Substantial economies are possible in U.S. energy input with the present structure of the economy, without sacrificing the continued growth of real incomes . . . . Our adaptation to a less energy-intensive economy would not reduce employment; in fact, it would result in a slight increase in demand for labor . . . . Other Project-sponsored studies also support the conclusion that we can safely uncouple energy and economic growth rates."

"First we can plug leaks and use thriftier technologies to produce exactly the same output of goods and services," says Lovins in his seminal book, *Soft Energy Paths.*

Energy efficiency measures range from better insulation to weather-stripping, more efficient furnaces and appliances, automatic flue dampers, set-back heating, to storm windows and doors; they include heat exchangers, smaller cars, less overlighting, recycling of materials, the use of waste heat, and on and on.

"Saving energy," said former Federal Energy Administration Director Frank Zarb, "is synonymous with saving dollars and can, in fact, be one of the least expensive energy supplies this nation has." He added: "Contrary to myth, conservation is vital to our efforts to sustain our high standard of living and rekindle economic growth."

As to why there has not been a stress on energy-efficiency up to now, Lovins speaks of a maze of "institutional barriers, codes, an innovation-resistant building industry, lack of mechanisms to ease the transition from kinds of work that we no longer need to kinds we do need, opposition by strong unions to schemes that would transfer jobs from their members to large numbers of less 'skilled workers', promotional utility rate structures, fee structures giving building engineers a fixed percentage of prices of heating and cooling equipment they install, inappropriate tax and mortgage policies, conflicting signals to consumers, misallocations of conservation's costs and benefits, imperfect access to capital markets, fragmentation of government responsibility."

"Wasteful energy consumption has been encouraged to increase the profits of energy-producing companies," is the way "Jobs and Energy"* sums it up.

After energy use is optimized through energy efficiency—and this alone, says Hayes, can allow the United States to meet all its new energy needs for the next quarter century—then come the other components of the soft energy path:

*Environmentalists for Full Employment, 1977.

SOLAR ENERGY

Solar energy "heads the list" for the soft energy path, stresses Lovins. As a 1976 report by the U.S. Energy Research and Development Agency declared: "Solar energy is the one source of energy for which there are no fundamental obstacles, no insurmountable barriers, no serious environmental problems." The Oregon Energy Council has declared: "A transition to a solar energy economy is desirable and realizable. It involves neither privation nor social deprivation . . . . The rewards would be enormous. Our children would have a totally indigenous, permanent, safe energy system which could be relied on by countless generations for the future." As U.S. Senator Charles Percy has said: "Solar energy is not an exotic dream of the future. Rather, it is workable today."

As the U.S. Council on Environmental Quality declared in 1978: "Our conclusion is that with a strong national commitment to accelerated solar development and use, it should be possible to derive a quarter of U.S. energy from solar by the year 2000. For the year 2020 and beyond, it is now possible to speak hopefully, and unblushingly, of the United States becoming a solar society."

"Even in the least favorable parts of the continental United States, far more sunlight falls on a typical building than is required to heat and cool it," says Lovins.

Two forms of solar energy are "passive" and "active."

Passive solar use involves the designing and building of structures to take maximum advantage of the sun–large windows facing south that can absorb and store the sun's heat, for example.

"Passive systems store energy right where sunlight impinges on the building's structural mass," explains Hayes.

Lovins calculates that if new buildings constructed in the U.S. in the next twelve years were built "properly" to take "advantage of passive solar" heat, "we could save about as much energy as we expect to recover from the Alaskan North Slope."

Active solar systems, in contrast, involve fans or pumps moving air or liquid from a solar collector to a storage area. Erecting a solar collector requires the basic skills of plumbing and once the collector is built, the fuel is free.

The development of the flat-plate collector to catch heat is credited, Hayes notes, to an 18th century Swiss scientist, Nicolas de Saussure, "who obtained temperatures over 87 degrees Centigrade using a simple wooden box with a black bottom and a glass top." The principle of a solar collector does not involve the outside temperature but the long wave lengths of sunlight which become trapped in a collector. This is why upon returning to a car left closed on even a mild day, you will find the car hot inside. The sun's rays have entered through the car's windows and the long wave rays have remained trapped inside the car.

The principle is simple. The technology is readily available.

Solar collectors can be placed beside a building, on its roof or recessed into the roof. They are ideal for heating space or water to temperatures up to 100°F. And this comprises the main part of the need for heat which is itself the main part of energy use. The heat is stored in rocks or salts.

More than two million solar collectors are in use in Japan and 200,000 in Israel. In northern Australia, solar water heaters are required by law on new buildings and they were widely used in California and Florida until the advent of cheap natural gas.

Solar collectors are being used for all sizes of structures. The town of Mejannes-le-Clap in France has embarked on a program to get the heat for the entire town from the sun. What is to be the largest solar-heated building in the world, a 325,000 square foot structure, is currently under construction in Saudi Arabia. (The solar energy reaching the Saudi Arabian desert each year equals the world's entire reserves of oil, gas and coal.)

Air conditioning with solar energy is being developed in Japan and the United States. "Fortuitously," notes Hayes, "solar air conditioners reach peak cooling capacity when the sun burns brightest, which is when they are most needed. Consequently, solar air conditioners could reduce peak demands on many electrical power grids."

Then there are photovoltaic cells. Fashioned from silicon, the second most abundant element in the earth's crust, they turn sunlight directly into electricity. Photovoltaics, the principal power source of space satellites, "have no moving parts, consume no fuel, produce no pollution, operate at environmental temperatures, have long lifetimes, require little maintenance," notes Hayes.

Photovoltaics can be placed on the roofs of buildings, eliminating transmission and storage problems.

Said the House of Representatives Committee on Government Operations about photovoltaics:

> Photovoltaic prices have already been reduced enormously. In 1973, a photovoltaically-generated peak watt of electricity cost about $300. In 1977, Dr. Henry Marvin, Director of ERDA's Division of Solar Energy, said the same peak watt could be generated for $15 to

$25.[295] In addition, by 1980, silicon crystal arrays, one of the more promising materials, are expected to produce electricity directly from the sun for $1 to $2 per peak watt—less than one-tenth of today's cost. Ultimately, Federal plans call for photovoltaic power to cost from 10 to 30 cents per peak watt by the year 2000.[296]

Solar cells, the FEA report says, could provide enough electricity economically to power street lights, light parking lots and airport runways and run irrigation pumps. The cells could even begin to meet some household electricity needs, and become a major energy source in developing countries in the next 5 years.[297]

Further, solar power could be generated at the site where it is needed, eliminating the need for new massive generating facilities and long distance transmission lines.[298] The savings could be substantial. Transmission and distribution now account for about 70 percent of the cost of providing electricity to the average U.S. residence.[299]

It is more advantageous to produce energy where it is needed rather than at large, distant powerplants. If on-site devices are used, it would be unnecessary to build solar systems in the image of huge nuclear and coal powerplants.

Hayes concludes: "Sunlight is abundant, dependable, and free. With some minor fluctuations, the sun has been bestowing its bounty on the earth for more than four billion years, and it is expected to continue to do so for several billion more. The sun's inconsistency is seasonal and reasonable, not arbitrary or political, and it can therefore be anticipated and planned for."

WIND POWER

The wind is also the product of the sun—generated by uneven heating of the spinning planet.

Says Hayes: "The wind power available at prime sites could produce several times more electricity than is currently generated from all sources." Some six million windmills were built in America over the last 100 years. Some 150,000 still spin productively. In the 1920's windmills were a major source of electrical power in the U.S. In the 1940's, a windmill erected on a hilltop near Rutland, Vermont generated 1.25 megawatts. A two megawatt windmill, the largest windmill ever constructed, has recently been built in Tvind, Denmark. It was put up cooperatively, for $600,000, for the Tvind school, with teachers contributing their salaries and with much of the labor donated. It provides all the school's energy needs—electricity and heat—and surplus electricity is sold to a local power company. In the Valley of Lasithi on Crete, there are 10,000—yes, 10,000—windmills pumping irrigation water.

The World Metereological Organization, Hayes notes, estimates "that 20 million megawatts of wind power can be commercially tapped at the choicest sites around the world, not including the possible contributions from large clusters of windmills at sea."

Against the current total world generating capacity of 1.5 million megawatts, wind can make a major contribution.

Radio station KMFU, powered only by a windmill, broadcasts from northwest Colorado. "The sound of the wind," it tells its listeners.

Home wind turbines are rapidly coming onto the market. In harvesting the power of the wind, big is not necessary. Indeed, smaller windmills can twirl and produce energy in much lower winds, can be readily mass-produced and can provide for a more decentralized pattern of energy generation.

Wind energy can be stored in batteries, or their mechanical motion can be used to compress air for power on a windless day. They can also, through electrolysis, break water down into hydrogen and oxygen and the hydrogen gas can be liquified or compressed and stored.

Simple windmills can be made cheaply. The Brace Research Institute of Canada has designed a windmill-water pump that can be made out of two halves of a forty-five gallon oil drum. It cost $50 to make and will operate at winds as low as eight m.p.h.

A series of windmills can provide energy to a small community. And for some communities, one wind turbine will do: some sixty per cent of the electrical needs of the 600 full-time residents of Block Island, in the Atlantic off Rhode Island, are supplied now by one recently-erected windmill.

A group of people in New York City's East Village had to battle with Con Edison to construct a windmill on top of an apartment building to supply them with electricity. But they won and the tenement turbine now freely spins.

Significant use of wind power in America was phased out in the 1930's with the Rural Electrification Administration's drive to distribute centralized, fossil fuel-fired power.

Good potential siting areas for wind turbines include the Great Plains, the Great Lakes, the Gulf and New England Coasts. Dr. William Heronemus, professor of civil and electrical engineering at the University of Massachusetts, projects having wind stations floating in the Atlantic off the New England coast with a total capacity of 82,000 megawatts—more than enough to supply the six New England states with the amount of electricity they now use.

HYDROPOWER

"Only a fraction of the world's hydropower capacity has been tapped," notes Hayes. By "the most conservative standards, potential hydropower developments definitely exceed one million megawatts," he says. Some surveys, he adds, suggest a potential up to three million megawatts. The current world hydroelectric capacity is 340,000 megawatts.

Hydropower was used by the Romans to grind grain. By the 1700's the water wheel was well-developed, and in 1882 the first hydroelectric facility started up in Appleton, Wisconsin.

By 1925, hydropower accounted for forty per cent of the world's electric power, Hayes notes.

Africa, Asia and Latin America are rich in undeveloped hydropower. Africa has twenty-two per cent of the world's potential but produces only two per cent of the world's hydroelectricity; Asia has twenty-seven per cent of the potential, produces twelve per cent; Latin America, with twenty per cent of the potential generates but six per cent.

Here again, big is not necessarily best. Hydropower through huge dams and extensive reservoir systems can easily wreak environmental havoc. Switzerland, Sweden and China have stressed smaller installations. "A small amount of water dropping from a great height can produce as much power as a large amount of water falling a shorter distance," notes Hayes.

To generate hydropower, smaller amounts of water falling from shorter heights can be fine, too: that's the basis of the "low head" dam. Low head dams, like the windmill, were forced out by cheap fossil fuels, and sites for them exist all over America.

Now "microhydroelectric" projects are coming into existence in many of the thousands of abandoned low head dam sites.

Water doesn't even have to fall to generate electricity. Electricity can be generated along rivers running at their normal level without dams—a "run of the river" process.

Other forms of hydropower:

Tidal Power. The English used power from the tides to mill grain at Bromley-by-Row in the year 1100. And a tide mill at Woodbridge, England, built in 1170 functioned for the next 800 years. In 1966, the French built the first commercial tidal energy facility, the Saint Malo plant on the Rance River, with a capacity of 240 megawatts. Excellent sites for tidal power have been identified off the coast of twenty-three countries. The French are considering a 6,000 megawatt plant on the Bay of Mont-Saint-Michel. The Russians have built a plant at Kislaya Guba and there is interest in the U.S. and in Canada again in taking advantage of the plummeting tides of the Bay of Fundy.

Ocean Thermal. Three quarters of the world is covered by ocean and a swatch of water about twenty-five degrees latitude on both sides of the equator captures large amounts of solar heat. Processes are being worked on to use both varied ocean temperatures and ocean water heat to turn turbines.

CO–GENERATION

Some forty per cent of the nation's energy goes to industry, much of it vented away as waste heat after use in industrial processes. By employing co-generation, this waste heat can be recycled to turn turbines to make electricity. Also the heat which otherwise escapes from chimneys can heat space when piped through a factory or to adjoining homes and businesses.

"If electric generation took place inside factories instead of at remote power plants," says Hayes, "the waste heat could be efficiently cascaded through multiple uses."

The recapturing of waste heat from industrial processes is widely used in Europe, but little used in America. In the 1920's and 1930's, major paper companies began producing electricity with steam produced for paper pulping. They found they could produce three to four times as much electricity as they needed, but the U.S. Justice Department took steps that forced the firms to choose between paper and energy production.

Co-generation has been estimated as saving one third to a half of the energy which would be required to separately run the systems needed to make steam and fire up industrial furnaces and those used to produce electricity and heat for industry. This combined use of energy would substantially reduce overall energy needs.

BIOMASS

"Green plants are nature's collectors of solar power. . . . All fossil fuels were once biomass," notes Hayes. But "unlike fossil fuels, botanical energy resources are renewable."

Projects Hayes: "As much energy could be obtained from biomass each year as fossil fuels currently provide."

Projects Lovins: "The whole of our transport needs could be met by organic conversion" of biomass.

Henry Ford was an early champion of alcohol fuel. He designed the Model T with an adjustable carburetor for use with alcohol. "Despite the intense competition from gasoline, alcohol fuels were used to power American cars well into the 1920's and 1930's," notes Ken Bossong of the Citizens Energy Project, "and even wider use was being made of alcohol fuels in other countries." Up to World War II forty nations were using alcohol fuels, Bossong has found, and the war prompted an even greater reliance on alcohol for propulsion. Hitler converted Germany's aircraft and other war machinery to run on alcohol fuels, with the destruction of Germany's refineries. The U.S. government had America's whiskey distilleries modified and alcohol was produced to fuel submarines and power torpedoes, and mixed with gasoline to run jeeps and aircraft. During the war years the U.S. production of alcohol increased six-fold to 600 million gallons of alcohol a year.

In 1917 Alexander Graham Bell said: "Coal and oil are going up and are strictly limited in quantity. We can take coal out of a mine but we can never put it back. We can draw oil from subterranean reservoirs but we can never refill them again. The world's annual consumption has become so enormous that we are now actually within measurable distance of the end of the supply. What shall we do when we have no more coal or oil? We can make alcohol—a beautifully clean and efficient fuel—from sawdust, the waste products of our mills . . . from cornstalks, and in fact from any vegetable matter capable of fermentation. Our growing crops and even weeds can be used. The waste products of our farms are available for this purpose, and even the garbage from our cities."

What Bell spoke of then is all still available now.

Again, small appears most efficient.

"Smaller scale operations," says Gene Schroder, a Colorado farmer and alcohol producer, "can benefit from the economies of integration. The alcohol facility is not just a liquid fuel producer, but a feed enrichment plant. Dewatered distillers' mash can be fed directly into a feedlot, or dairy, hog, poultry, or fish-farming operation. Animal waste can be processed through an anaerobic digester to produce fertilizers, feed supplements and methane . . . . The by-product fertilizer goes back on the soil . . . . It can be argued that such integrated systems work best on a smaller scale."

Attention is now being focused on the growing of "energy crops." These range from winter wheat to good old wood to water hyacinths to a variety of shrubs including *Euphorbia lathrus* and *Euphorbia tirncalli* "whose sap," says Hayes, "contains an emulsion of hydrocarbons in water." Nobel laureate Melvin Calvin estimates such plants could produce ten to fifty barrels of oil per acre per year. Both thrive on dry, marginal land.*

Energy efficiency in transportation–small cars, not gas guzzlers, an emphasis on mass transit–will aid in allowing alcohol to fill the vehicular fuel need. Much of this efficiency will come about from what Lovins calls "increased energy productivity driven by economic rationality."

Explains Hayes: "Abandoning automatic transmissions would save one-tenth of automotive fuel use. Switching to radial tires would save another tenth. Since fuel consumption decreases about 2.8 per cent for each 100 pounds of weight reduction, reducing the size of the average American vehicle from 3,600 pounds to 2,700 pounds would save one quarter of the United States' present gasoline use. A further reduction to 1,800 pounds would reduce automobile fuel needs by nearly half. These smaller cars would require smaller engines, which would cut fuel requirements still more."

Says Hayes: "Photosynthetic fuels can contribute significantly to the world's commercial energy supply . . . . Plant power can, without question, provide a large source of safe, lower-polluting, relatively inexpensive energy."

As to electric cars, Lovins sees "no evidence" that they "could compete with the present best practice in far more efficient fueled cars" operating on alcohol "derived with present technology from biomass residues."

Also a component of biomass:

Power From Waste. U.S. waste, estimates Hayes, could provide for seven per cent of the American energy budget. Power plants fired by refuse have been busy in Europe since before the turn of the century and are now beginning to spread throughout America. Further, waste can be converted to "biogas," including methane, and through the pyrolysis process into oil. Recycling of waste should go hand in hand with energy production from it. "The American trash heap," notes Hayes, "grows annually by more than 11 million tons of iron and steel, 800,000 tons of aluminum, 500,000 tons of other metals, 13 million tons of glass, and 60 million tons of paper; some 17 billion cans, 38 billion bottles and jars, 6 million discarded television sets and 7 million junked cars and trucks contribute to the total. The en-

*"Hydrocarbons via Photosynthesis," a paper presented before the American Chemical Society, San Francisco, October 5-8, 1976.

ergy required to produce a ton of steel from urban waste—including separation, transportation and processing—is only 14 per cent of that needed to produce a ton of steel from raw ore. For copper, the figure is about 9 per cent, for aluminum only 5 per cent."

As well, a study conducted by the Ford Foundation's Energy Policy Project committee found that if the American paper industry were to use its wood waste as fuel, it would reduce its fossil fuel consumption seventy-five per cent.

GEOTHERMAL

The earth's core is molten hot. The heat can be captured and converted into energy all over the planet. Wilson Clark, in his key book on energy options, *Energy for Survival, The Alternative to Extinction,** estimates that "total energy . . . from geothermal resources is virtually limitless . . . many times more energy than the world could ever use."

United Nations energy consultants John Banwell and Dr. Tsvi Meidav have estimated that the geothermal energy stored in the upper 24,600 feet of the earth's crust is equivalent to twenty-one million tons of oil per square kilometer of earth's surface.**

The heat of the inner earth breaks out in spots where the molten core has moved close to the surface: a vòlcano is one manifestation, a hot spring a gentle form. Between the two are a myriad of areas which produce steam and/or hot water under pressure, or can do so.

Geothermal energy is widely tapped in Iceland (the city of Rejkavik gets ninety per cent of its heat from geothermal sources), New Zealand (seven per cent of the country's electricity is from geothermal power), Italy (the first use of geothermal power for making electricity began in Landarello in 1904), and in the U.S., particularly its West. Dr. Robert Rex and his colleagues at the Institute of Geophysics at Riverside, California estimate that the geothermal energy potential stored in the Imperial Valley area of California equals up to sixty-five per cent of the thermal capacity of the entire world's oil reserves.

As "strictly a transitional fuel," emphasizes Hayes, is plentiful and abundant coal, which he and others stress should be used with new combustion technologies, including fluidized beds and anti-pollution scrubbers, to reduce its environmental consequences.

*Anchor Press, Garden City, New York, 1975.
**"Geothermal Energy for the Future," a paper presented at the Annual Meeting of the American Association for the Advancement of Science in Philadelphia, December 1971.

"Once we do understand the energy crisis," says Barry Commoner, "it becomes clear that the nation is not poor, but mismanaged; that energy is not wasted carelessly, but by design; that the energy we need is not running out, but is replenished with every dawn; that by relying on our solar resources we can foreswear the suicidal prospect of a war that would begin with oil but end with a nuclear holocaust. The solution to the energy crisis—the solar transition—is an opportunity to turn this knowledge into action, to embark on a new historic passage."

Or as Lovins concludes in his work, *Is Nuclear Power Necessary?*— "Nuclear power is not necessary and indeed is a positive encumbrance whose resource, political and infrastructural commitments get in the way of what we should be doing instead."*

One strength of the soft energy path is its flexibility. The soft path "minimizes the economic risks to capital in case of error, accident or sabotage," notes Lovins, while "the hard path effectively maximizes those risks by relying on vulnerable high technology devices, each costing more than the endowment of Harvard University." The soft path is "more flexible—and thus robust. Its technical diversity, adaptability and geographic dispersion make it resilient and offers a good prospect of stability under a wide range of conditions, foreseen or not."

"The hard path, however," Lovins continues, "is brittle. It must fail, with widespread and serious disruption, if any of its existing technical and social conditions is not satisfied continuously and indefinitely."

Another strength of the soft path is its decentralized nature. The soft path is about peoples' technology: people in control of their energy. As Stuart Diamond and Paul S. Lorris write in *It's In Your Power,* "We have allowed ourselves to be enslaved by energy, and, by extension, by the people who have provided it. We have trusted the energy companies. We have trusted government officials to protect us" and "those people in our civilization who were supposed to handle energy—government, oil companies, and other large corporations—have not done their jobs . . . . We are now at a crossroads. We can control the power we use, producing it ourselves or forcing power companies to follow our wishes. Or we can continue letting those who do not have our interests at heart make decisions for us, and accept the results. If we choose the latter, we are gambling with our future, trust-

*"Energy Path No. 3, Friends of the Earth, London, 1979.

ing those who have failed us many times before. If we choose the former, we are accepting responsibility for our own destiny—and the reward will be the quality of life that has been just beyond our grasp. A new day dawns. We open our eyes and see the sun rising. It looks beautiful there in the sky, radiating comforting warmth and power. We should embrace it now in all its many forms."*

What about fusion?

This has been held out by the nuclear establishment as a somewhat cleaner form of nuclear power—as the hydrogen bomb, a fusion device, is somewhat cleaner in fall-out than an atomic bomb. Somewhat.

Fusion is theoretically supposed to get its power from fusing nuclei together. This would be the opposite of fission, which blasts the nuclei apart. But to start the process, extremely high temperatures are required—100 million degrees Centigrade, more than six times the estimated temperature of the sun's interior.

Although Dwight Eisenhower, when he was President, suggested that the AEC keep the public "confused about fission and fusion,"** fusion is a dirty, radioactive process, too.

The theory is to fuse deuterium and tritium atoms. Large amounts of tritium would be used. Tritium is highly radioactive. At the temperature a fusion reactor would operate, matter exists only in gaseous "plasma" which walls, as we know them, cannot hold. Containment of tritium is seen as impossible.

Hundreds of tons of radioactive waste would be produced annually. Further, a fusion reactor's "fuel supply would not be limitless," stresses Hayes. "Tritium is derived from lithium, an element not much more abundant than uranium." He adds that "the intense radioactivity of the equipment would make maintenance almost impossible."

Neutron activation along with radioactivity from tritium would be major sources of radioactivity in a fusion reactor. This neutron activation is the reason the fusion reactor is eyed as a fuel factory for nu-

*Rawson, Wade Publishers, Inc., New York, 1978.

**From classified AEC documents, disclosed during U.S Congressional hearings led by Senator Edward Kennedy in April 1979 on the federal government's responsibility for cancers caused by the testing of atomic weapons. Gordon Dean, chairman of the AEC, declared in a May 17, 1953 memo after speaking to President Eisenhower: "The President says, 'keep them confused about fission and fusion.'"

clear weapons. There is a "hybrid" fusion-fission reactor now being pursued which would have uranium -238 placed around the plasma. With the neutron bombardment, massive amounts of plutonium would be produced.

What about the argument—"well, the rest of the world is doing it—shouldn't the U.S continue to move in the direction of nuclear power, too?"

This is Lovins' reply: "The genie is not wholly out of the bottle yet—thousands of reactors are planned for a few decades hence, tens of thousands thereafter—and the cork sits unnoticed in our hands . . . . The most important opportunity available to us stems from the fact that for at least the next five or ten years, while nuclear dependence and commitments are still reversible, all countries will continue to rely on the United States for the technical, the economic, and especially the political support they need to justify their own nuclear programs . . . . In almost all the countries the domestic political base to support nuclear power is not solid but shaky. However great their nuclear ambitions, other countries must still borrow that political support from the United States . . . . My own judgment based on the past ten years' residence in the midst of the European nuclear debate, is that nuclear power could not flourish there if the United States did not want it . . . . I am confident that the United States can still turn off the technology that it originated and deployed. By rebottling that genie we could all move to energy and foreign policies that our grandchildren can live with. No more important step could be taken toward revitalizing the American dream and making its highest ideals a global reality."

CHAPTER NINE

What You Can Do About It

What *can* be done about nuclear power?

One thing you cannot do is move away from it. Nuclear plants and elements of the "nuclear cycle"–from uranium mines to mountains of mill tailings to enrichment plants to waste dumps–proliferate. And in the few areas free of them, nuclear weapons are produced or deployed.

Here is a map of the picture in the U.S.:

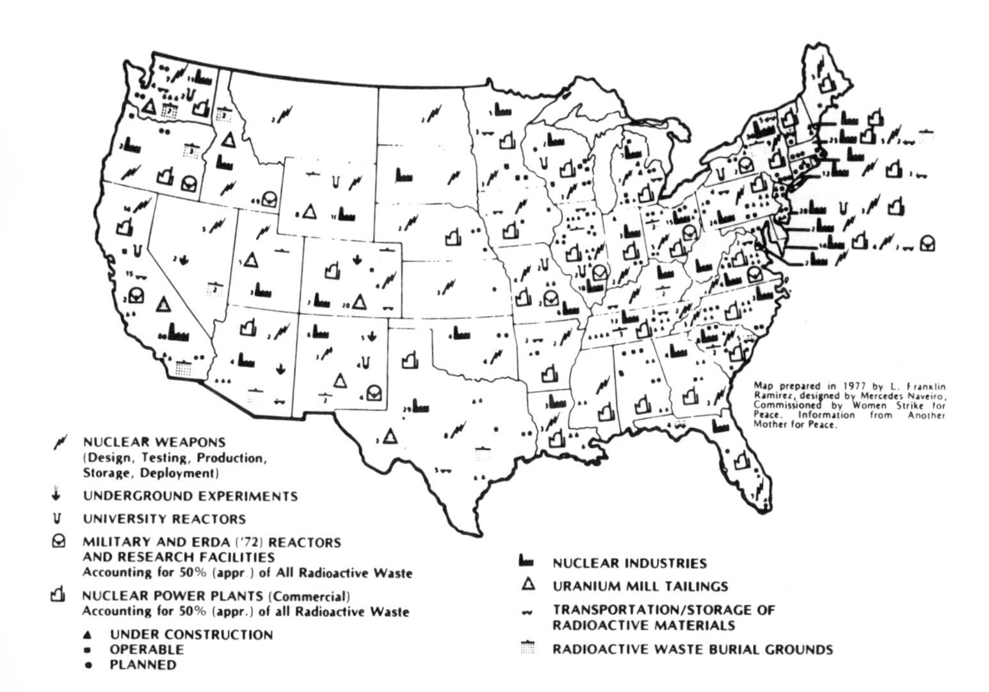

There is no escape to sea.

In Jacksonville, Florida today a subsidiary of Westinghouse, Offshore Power Systems, is designing and preparing to build floating nuclear plants—which it calls FNP's—for placement offshore.

Here's what one is supposed to look like:

The idea for floating nuclear plants is credited to Richard Eckert, a vice president of the Public Service Electric and Gas Co. of New Jersey, and came to him, according to the utility, while he was taking a shower in 1969. In the shower, Eckert thought the sea could supply the mammoth amounts of water nuclear plants need. The utility got Westinghouse to agree to manufacture them.

For $180 million, Offshore Power Systems has built an "FNP manufacturing facility" on 875 acre Blount Island. The firm hopes to be licensed by the NRC to start fabricating eight of these plants in late 1980 or 1981.

Betsy Molloy, chairperson of the Atlantic County Nuclear Advisory Committee in New Jersey, one of the groups which have been fighting the licensing of the "FNP's" speaks of them "contaminating the ocean." The EPA cites accident "factors unique to offshore siting," particularly "molten core-sea water steam explosions."

But the U.S. Nuclear Regulatory Commission, like the AEC (which it replaced), has never denied a license for any kind of nuclear plant anywhere.

The NRC's Office of Nuclear Reactor Regulation has already given clearance for the FNP's to be built.

No. 79-12
Contact: John Kopeck
Tel. 301/492-7715

FOR IMMEDIATE RELEASE
(Mailed - January 11, 1979)

NRC ISSUES FINAL ENVIRONMENTAL STATEMENT
ON FLOATING NUCLEAR POWER PLANT PROJECT

The Nuclear Regulatory Commission's Office of Nuclear Reactor Regulation has issued a third and final part of Final Environmental Statement on the environmental considerations associated with the application of Offshore Power Systems, Inc. of Jacksonville, Florida, for a license to manufacture eight floating nuclear power plants.

Offshore Power Systems would manufacture the floating nuclear plants and test them at Blount Island, Jacksonville. Each plant would use a pressurized water reactor and would have a net electrical capacity of about 1150 megawatts.

If specific sites designated by purchasers of the floating nuclear plants are approved by the NRC, the plants involved would be towed to and moored at such sites, and nuclear fuel would then be loaded into the reactors.

The final statement provides a comparative assessment of the overall risks between floating and land-based nuclear plants relative to a spectrum of accidents, including a core melt, which involve the release of radioactive materials to the environment. The statement also presents a cost-benefit analysis section and summary and conclusions which reflect the overall environmental review conducted by the staff and documented by the three-part environmental statement.

The NRC staff has concluded that, from the standpoint of environmental effects and subject to certain conditions, the manufacturing license should be issued. The conclusion is based on a weighing of the environmental, economic, technical and other benfits against environmental costs and available alternatives.

The NRC staff also concludes that the proposed plants can, with suitable modifications, be sited and operated as electric generating stations in the nearshore and offshore waters of the Atlantic Ocean and the Gulf of Mexico and at carefully selected and appropriately modified estuarine and riverine locations.

Offshore Power Systems gives this sales promotion for floating nuclear power plants:

FNPs and the future.

The demand for electricity in the United States, which fell off following the oil embargo of 1973, is on the rise again. Today, one quarter of the energy we consume is electricity. By the 1990s, this will increase to almost one-half.

How will the demand be met in the years ahead?

By 1990, it is estimated that about one-third of the country's electricity will, of necessity, be generated with nuclear power.

Floating Nuclear Plants will be an important part of that total.

FNPs offer utilities unique siting flexibility. They offer shortened production schedules, improved product quality, reduced environmental effects and lower costs. And they offer a licensing procedure with a distinct advantage over other nuclear plants.

Floating Nuclear Plants from Offshore Power Systems are an innovative answer to the country's energy problems.

Meanwhile, there is the plan for immense "nuclear energy centers."

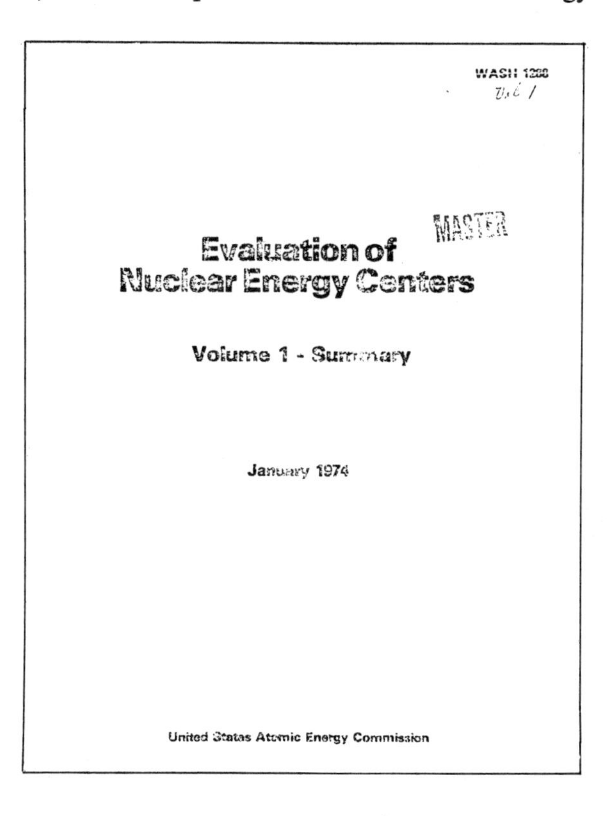

WASH 1288
Vol 1

MASTER

Evaluation of Nuclear Energy Centers

Volume 1 - Summary

January 1974

United States Atomic Energy Commission

These would take up as much as fifty to sixty square miles, according to the government. Sites have already been evaluated.

Says the government:

<u>Technological Aspects</u>

4.4.1 The basic nuclear energy center concept can be implemented using technologies now existing or available by the time of need. LWRs and HTGRs in the sizes now available will be the types of reactors used initially. However, during the 1990-2000 period, it should be possible to introduce breeders into the center and perhaps larger single unit capacities.

4.5.2 Nuclear energy centers provide increased opportunities for the innovation of new construction practices through standardization of design and modularization of components. Modification of traditional construction practices which offer the potential for reducing construction costs and lead times on nuclear plants thus have been considered. GE has developed one such plan which involves the simultaneous construction of at least four plants on an overlapping schedule. Although nuclear energy centers are not necessary for the implementation of this plan, the plan is compatible with the development of such centers. The plan has not been demonstrated and there are many problem areas to be worked out but it merits further consideration for incorporating into nuclear energy centers. Offshore siting of nuclear energy centers also offers the possibility of innovative construction practices which would result in lower costs and faster construction.

And the government describes its promotional role:

Government's Role

The government's role in nuclear energy centers can be divided into many categories: studies and evaluations, research and development, site selection, regulatory actions, and promotional activities. The government should institute programs to define and resolve issues related to center vulnerability, materials diversion, insurance, environmental effects, socio-economic impact and regulatory requirements. In the area of research and development, possible technical constraints need to be assessed and the development of siting guides and criteria is necessary. If the concept is to have significant impact, the government needs to encourage or participate in the acquisition of an inventory of suitable suites. The government should promote the nuclear energy center through continued contacts with utilities and other involved organizations, policy statements encouraging this concept and public information.

At the same time, nuclear war—the other head of nuclear power—is regarded as feasible. Edward Teller, "father" of the hydrogen bomb, long-time chairman of the AEC's Committee for Reactor Safeguards and advisor to the Rockefeller family on nuclear matters, faces the possibility without shrinking: "Properly defended we can survive a nuclear attack," he says. There is "no doubt" that millions of people would die, he concedes, but "most people" can be saved, he says. He proposes that bulldozers be used to push aside debris and topsoil saturated with radioactivity and outlines ways of storing food and reconstructing factories. He says the "best guarantee" of a "postwar" source of energy would be construction of underground nuclear power plants. Teller speaks of survivors emerging "from their shelters" into a "kind of world man has never known." Still, with adequate preparation and organization after an "all-out war" the United States could, says Teller, "re-establish economic strength sooner than Russia —and so the United States would remain by far the strongest nation in the world."*

And Dr. Weinberg, now director of the heavily government-funded Institute for Energy Analysis is calling for "a cadre that from now on can be counted upon to understand nuclear technology, to control it."

"Fears about fission energy will subside as the public acquires familiarity with radiation," he says. Further, says Dr. Weinberg, "If a cure for cancer is found, the problem of radiation standards disappears."

"We're in the hands of lunatics and at the crossroads of time," Dr. Helen Caldicott has said. "It's time we rise up and say 'this is our world, we want to live.'"

"The splitting of the atom moved man into a new dimension of violence," Robert Jungk has written. "What began as a weapon against one's enemies now threatens the hand that wields it. In substance there is no difference between 'Atoms for War' and 'Atoms for Peace'. Nothing, neither stated intentions nor applied technology, can alter the inherent dangers of nuclear energy." Nuclear power, writes Jungk, "now looms like a great dark shadow over mankind. It is there, whether we see it as an inescapable blight or as the ultimate destruction of all life on this planet."**

A small group subjugating the rest of mankind, causing injury and misery and death, is not new on this earth. What is new since man unleashed the dreadful atomic genie is, as Einstein has noted, "the scale" of things.

Through much of world history, Lewis Mumford has written, there have been authoritarian and democratic cultures existing "side by side." With "the knitting together of a scientific ideology," the authoritarian structure has re-emerged on an unprecedented scale. "The inventors of nuclear bombs, space rockets, and computers are the pyramid builders of our own age: psychologically inflated by a similar myth of unqualified power, boasting through their science of their increasing omnipotence, if not omniscience, moved by obsessions and compulsions no less irrational than those of the earlier absolute systems: particularly the notion that the system itself must be ex-

Legacy of Hiroshima, Doubleday & Company, Garden City, N.Y., 1962.
**The New Tyranny*, Grosset & Dunlap, Inc., New York, 1979.

panded, at whatever eventual cost to life."

Democracy "consists in giving final authority to the whole, rather than the part," continues Mumford, who fears "we are now rapidly approaching a point at which, unless we radically alter our present course," democracy will be suppressed "or will be permitted only as a playful device of government, like national balloting for already chosen leaders in totalitarian countries." And it will be a "technical and managerial elite" in charge, "the sacred priesthood of science, who alone have access to the secret knowledge by means of which total control is now swiftly being effected." They are offering a would-be "magnificent bribe" of material advantage in return for "surrender" to a lethal machine, a machine which "deliberately eliminates the whole human personality, ignores the historical process, overplays the role of the abstract intelligence, and makes control over physical nature, ultimately control over man himself, the chief purpose of existence . . . . We must return to the human center . . . . We must ask, not what is good for science or technology, still less what is good for General Motors or Union Carbide or IBM or the Pentagon, but what is good for man: not machine-conditioned, system-regulated, mass man, but man in person moving freely over every area of life."*

Mumford adds: "If we focus our attention on now available solar energy, we shall find that all our basic energy needs are at hand in abundance . . . the key to having a sufficient supply of energy is to detach one productive process and function after another from the corporate power network, and restore them to the identifiable human communities capable of actively utilizing sun power and plant power, man power and mind power, instead of surrendering all authority to machines, mechanical organizations and electronic computers and in the end to their ultimate monitors and rulers, the Power Elite."**

The Trilateral Commission's *The Crisis of Democracy* concludes that "the problems of governance in the United States today stems from an excess of democracy . . . . Needed, instead, is a greater degree of moderation in democracy . . . . Democracy is only one way of constituting authority, and it is not necessarily a universally applicable one. In many situations the claims of expertise, seniority, experience, and special talents may override the claims of democracy as a way of constituting authority."

*From a speech by Mumford, a social philosopher, at the Fund for the Republic Tenth Anniversary Convocation on "Challenges to Democracy in the Next Decade," held in New York City in January 1963.

**From an address by Mumford in 1974 to the MIT Technology and Culture Seminar on "Enough Energy for Life."

Today democracy is threatened—on a vast scale. Directly connected to this: human survival is threatened—on a vast scale. This is new under the sun. *The scale.*

The Tellers and the Weinbergs, the G.E.'s and the NRC's, the Westinghouses and the Chase Manhattan Banks, Standard Oil and the Gettys, this authoritarian establishment barren of morality, crazed with power, in worship of nuclear power, is using in nuclear power an instrument which can kill not just thousands. It can kill us all. And unless it is stopped, it very well may.

To stop it will take exactly what it wants to eliminate in order to perpetuate itself: democracy.

If people are told the truth and can choose, then there is a chance. People will bump through history, but in the collective wisdom of democracy they will not, time has shown, kill themselves en masse. Only despots cause this.

Today we are in the hands of technological despots and the threat is unprecedented. *The scale.* Will it be a breeder reactor in Russia or France or Japan going up? Will it be the carnage of nuclear war as little nations convert their nuclear fuel into atomic bombs? Will it be a miscalculation in nuclear confrontation between the current "nuclear nations?" Will it be a nuclear runaway spewing fission products from a light water reactor near Chicago, Los Angeles, Paris, London, New York—bathing your town in radioactivity? We have been thrown into a mad, Faustian bargain without our vote, without our permission, without our being given the facts. A few now threaten the future of us all.

It is time this all be unmasked. It is time it is stopped. No more cover ups. The death trip of nuclear madness must be undone, and it must be undone now.

Whatever we do means nothing if man cannot survive.

Oppenheimer saw it that July day:

I am become death,
The shatterer of worlds.

This need not be. We need not be sheep led to a slaughter. We can survive. In democracy. In humanity. The billions of dollars, the entrenched interests can be overcome. There can be a future—if we *will* it.

This mobilization for survival will require every kind of action. The nuclear industrial establishment is so huge and powerful that no one way will work.

Here is how people from around the world believe nuclear power can best be stopped.

WARREN LIEBOLD
Sierra Club, U.S.A.

I think the safe energy movement has two basic tasks: stopping nuclear power and building an energy infrastructure based on the elegant use of renewable sources of energy. Different people have different preferences for tactics. We need to intervene in the licensing process to get the NRC to consider its own inconsistencies and bias. We must fight in the political arena against the industry lobbyists who effectively control Congress and many state legislatures. We must non-violently place our bodies in the way of nuclear construction. People who choose one tactic must respect those who have chosen other tasks because we have no time to lose in in-fighting. Building the alternatives is the flip side of the coin and no less important than stopping nuclear power. Building codes and discriminatory utility rates must be reformed and federal energy policies must be re-ordered toward solar development. Small hydro sites must be wrested from the utilities which won't develop them and the legal barriers which utilities have erected against co-generation development must be overcome. We will have failed if the solution to the energy problem ends up making the rich richer and the poor poorer. In addition, the alternatives should be built by example. Community recycling and insulation projects, well-made solar systems and well-sited windmills are physical proof that there is another way and can be put into action long before Washington does much of anything. On top of all these activities must be constant education both of ourselves and of the American public of which we are a part. Day in and day out the energy industry's lies and scare campaigns must be countered with a positive and accurate safe energy analysis. In the end our task is to convince the American people that we don't have to make a pact with the devil to solve the energy crisis and that we must work toward that solution with all due speed.

DAVID BROWER
Friends of the Earth, U.S.A.

The present era of nuclear roulette poses a far greater threat to all living things than the Vietnam war did, because proliferation could stumble us into the final war. The Vietnam war and the subsequent proliferation were triggered by the deepening global addiction to exponential growth in material wants, in energy consumption, and in the buildup of military-industrial strength of the nations competing to secure those wants—to preempt the resources essential for dominance. It is not too late to change course, but it will be too late too soon if too few urge the change. 1) Exponential growth has led to a destructive race for more and more energy. That race led to nuclear power, the once bright hope, which has turned into the greatest threat to the future—nuclear proliferation, which no one wants. 2) While deploring nuclear proliferation in words, the great powers are racing to exacerbate it. 3) If they succeed, the inevitable war will so damage the world's major civilizations that they can never recover. 4) The U.S., and probably only the U.S., can lead the world back from the nuclear brink to which it led the world with good intentions. 5) Instead of assuming such leadership, the U.S. is pushing reactor sales and, like Russia, getting even deeper into the nuclear arms race. Both assume that superior sales, armament, and intimidation mean surer security. 6) Each reactor sold to a non-nuclear nation, however, expedites that nation's development into atomic-weapons capability. Indira Gandhi proved it. 7) Adding to our nuclear capability while denying it to have-not nations won't work. They have reason not to trust the haves. Mistrust and inequity breed sabotage and terrorism. 8) If it is to lead others back from the brink, the U.S. must turn around itself, renouncing nuclear power as the first step; we can do so quickly. 9) Other nations will note, upon analysis, that by moving back also, they can stop wasting themselves on the costly nuclear experiment and invest instead in progress as if survival mattered. 10) Itself freed from the exhorbitant cost of going nuclear, the U.S. can allocate resources to the sustainable alternative, derived from the sun's constant gift to the earth. The U.S. and other nations can pursue this alternative at home and abroad to the benefit of all. 11) As its constant gift, the sun sends to earth every few days (estimates range from three to twenty-one) as much energy as there is in all the world's recoverable fossil fuels. The daily gift is stored in rock, water, wind, and living things. 12) The sun has proved its reliability for eons. The nuclear tinkerers cannot prove the same of their gear, and if their tinkering goes on another decade, nothing worth tinkering with is likely to be left. 13) Each day's delay in ending the nuclear experiment makes less likely the chance to put logical solar substitutes in place in time. 14) People are sorely needed now to work the solar side of the street during the next ten years. The opportunity is real, immediate, and is not likely to be offered again. 15) The world can then breathe easier. History will admire the brilliance that went into trying to get the atom to work peacefully, but will admire more the wisdom of choosing not the radioactive, but the sunlit path. The foregoing points evolved in my own thinking over the past nine years. I have been a conservationist for forty years, twenty-three as a staunch advocate of nuclear power. But I have become a born-again believer in the pleasanter and sustainable solar alternatives. They deserve all the genius we can give them.

SISTER MARIA AIDA VELASQUEZ
Philippines Movement for Environmental Protection

Nuclear power can best be stopped through effective and sustained information dissemination and friendly linkages among various concerned groups which can promote global action against nuclear power. People have to know and understand nuclear power and the dangers it poses. I feel that despite the local publicity that the Bataan nuclear plant got here, many Filipinos, among them professionals, still have to get a real grasp of the hazards of nuclear power to health and environment. The majority have a vague idea that nuclear power plants are unsafe, especially after the President declared an indefinite suspension of the nuclear project; however, the level of awareness is not enough to expect committed action from them, in general. Effective and sustained information dissemination and discussion in groups is needed. The majority have to see the tie-up between nuclear power and proliferation of nuclear weapons. Many Filipinos are not well-informed on the horrors of Hiroshima and Nagasaki. The problem of nuclear power has to be seen as a whole–it is a concrete case of the efforts of a few well-placed and powerful men in developed countries to perpetuate their control on the life of billions of people, both in the developing and developed countries. More and more people have to responsibly ask–What are the consequences of inappropriate technology, like nuclear power, in a developing country? What misery does it breed? How does big business push nuclear power, especially in developing countries? Is there a real need for more power consumption? Ultimately, what lifestyle will answer the needs of the masses of people of the world? All these presuppose growth in freedom of people to decide what affects their lives, and their desire for a global community.

TH. BUCK
The Scottish Campaign to Resist the Atomic Menace

Stopping nuclear power will require the development of a mass movement in Britain which not only follows traditional "constitutional" campaign methods but will also have to commit itself to widespread direct action including economic action by the trade unions. Sympathetic public opinion alone will not achieve our goal. We shall have to supplement what is achievable within the British parliamentary democracy with what is achievable outside it. In all our campaign methods we shall have to maintain a high level of imagination to provide the inspiration necessary to motivate people to action. Ultimately, what is required is a broad psychological change in British society, away from the philosophy of growth for growth's sake and towards an appreciation of the possibilities and practice of more harmonious ways of relating to our environment and to each other. This can only come about if the movement as a whole develops an alternative energy programme and ensures that emphasis is placed on this to avoid being labelled as merely "anti."

LORNA SALZMAN
Friends of the Earth, U.S.A.

The anti-nuclear struggle in the U.S.A. is a little more than a decade old, yet in that time the opposition, with little money and against great odds, has succeeded in not only putting the nuclear industry on the defensive but in slowing the U.S. reactor program to a near halt. Reactor manufacturers are relying on filling back orders. This came about because of citizen and community action on many fronts and on many levels: in administrative hearings, interventions, lawsuits, lobbying in the halls of Congress and state legislatures, national petitions to Congress, dissemination of public information. In recent years, more direct action in the form of marches, picketing, demonstrations and actual site occupation is being taken by people and groups impatient with the slowness of the political process. No one of these strategies can work by itself. An intervention consisting of technical cross-examination behind doors cannot work unless propped up with local political and media action, using the hearing process to get out the facts about the dangers of nuclear power which in turn educates the public and elected officials and marshals opinion against nuclear energy. Similarly, direct action may appeal to younger or more radical factions but can only work when the surrounding community has been sensitized to the issue and can understand the social, political and economic implications of nuclear power. Each of these strategies has its own constituency which together form a potential united force that politicians and media cannot ignore. It is important the issue be seen as universally threatening and that its implications be spelled out so the issue can cross racial, age, economic and social lines. Then it becomes a powerful force for further political change.

No one strategy will suffice; all must be mutually supportive. In the event that one fails, there will be others to fall back on. There is no one way to stop nuclear power but a combination that unites people across community lines. If you're talking to workers, tell them that nuclear power plants substitute energy for labor and deprive other sectors of the economy of capital that could be used to create jobs—and tell them workers are on the firing line for radiation hazards and reactor accidents. If you're talking to investors, tell them the marketplace is drying up and investors will also be left holding the bag when utilities experiencing accidents like Three Mile Island are forced to pay for repairs, liability and replacement power. If you're talking to women, children and religious groups, stress the immorality of mortgaging the future and the human gene pool for a little extra electricity that we don't really need at all. If you're talking to consumers and the poor, tell them how nukes raise electric rates because they cost more than other kinds of power plants and because utilities make money by building new plants, needed or not, not from selling power. If you're talking to economists, tell them that money invested into energy efficiency and alternative energy has a faster payback period, thus freeing up money sooner for reinvestment, and tell them that money poured into energy development that doesn't bring returns for ten or more years fuels inflation. Don't expect any one strategy to win; direct action like occupations, rate withholding and demonstrations keeps public consciousness and media awareness high, but it must be backed up by lobbying, legislation and interventions in state and federal proceedings. Those 500 occupiers will sooner or later be dragged off site to jail and then the siting board or the Nuclear Regulatory Commission will go ahead and license the plant to operate anyway. Use every political and technical tactic and argument available to construct a solid case. Become a one-issue voter and put it to politicians running for office. Make nuclear power the overriding political issue of the day. It is, you know.

DAVID GARRICK
The Canadian Scientific Pollution and
Environmental Control Society

The way we're trying to do it here involves three things: the first is showing there is a better way; the second, teaching people and learning ourselves that there is a greater purpose being here and in a sense living is finding our way within that greater purpose. I'm talking about ecological harmony and coming to realize that we're part of a sustaining ecosystem. The third is we have to discover what is threatening to the ecosystem and act on it locally by doing all three things. Your lifestyle starts to come in tune with what the real needs are and at the same time a way is found to bring people together to affect qualitative changes. In British Columbia, we're looking at entire communities becoming energy self-sufficient, where there's no need to rely on any continental energy grid, where in fact you utilize the local resources with respect: gardening the forests, making as much liquid fuels as we need from wood waste, tapping into the sun and all its manifestations, the winds, the tides, wherever appropriate.

IRVING LIKE
Environmental Attorney, U.S.A.

The Administration's nuclear power plant program will be abandoned if it becomes apparent that it does not command a domestic consensus and its cost is so prohibitive as to raise intolerable economic problems. There is an analogy here to Vietnam which justifies applying the label "technological Vietnam" to the nuclear fission technology. The Vietnam war was terminated by the U.S. government when political and social opposition destroyed any domestic consensus for continuing the war and when the government and major corporate power centers concluded that the cost of the war was threatening to wreck the economy. Thus, anti-nuclear strategy should focus on persuading government and corporate policy makers (particularly the financial markets) that the nuclear fission option threatens to destroy the political, social and economic fabric as did Vietnam. How to do this? Historically, major changes in national policy have occurred when the fact finding processes of two branches of government worked against the third to produce a new public consciousness that something was rotten. The civil rights movement was fostered by an activist Supreme Court and Executive branch working together to achieve legislative reform. The Watergate results were the culmination of investigations by the judicial branch and the Congress focused on wrongdoing in the Executive branch. Now again it is the Executive branch (DOE, NRC) and the corporate interests promoting the nuclear fuel cycle which should become the target of the combined processes of the legislative and judicial branches (federal and state). Legislative and grand jury investigations, interventions in administrative proceedings, lawsuits, testimony before legislative hearings, as well as the various forms of petition for redress of grievances—these are available and effective political tools. They should be used to gain an anti-nuclear and pro energy-conservation, solar consensus, and convince

Wall Street and those who make the key investment decisions that nuclear is not the way to go.

RICHARD LERCARI
SHAD Alliance, U.S.A.

I do not believe any one type of activity could end nuclear madness by itself. However, I am convinced that the collective efforts of citizens joining together on all fronts will force nukes to be abandoned. Ultimately the laws which protect nuclear studies, power and weapons will be changed. It may take a revolution to stop nuclear pollution. How to do this: self-education, be informed of the nuclear issue in general and see the connection with other struggles; sharing information and your feelings with relatives, friends and neighbors; speaking out publicly; supporting citizen action locally and nationally, with your time, energy and resources; joining with your neighbors in common projects of mutual strength; self-activity, take the initiative where and when needed, don't wait for someone to organize you; questioning; awareness of energy consumption; conservation and recycling on the job, at home, in the community. I choose the nonviolent direct action movement, which offers us methods of dealing with each other and the faceless technology in a humanizing and creative way, encouraging and empowering each other to take control of our lives and shape the future we will live in together. The Alliance, a network of mutual support regionally/nationally. The Local group, neighbors who are active and aid the community in self-education and mobilization—together. The Affinity group, a collective effort of citizens who share a common commitment to a project or action, often the basic unit of support and making decisions consentually in a large campaign. Participatory decision-making. Consensus, not leaders and followers. Being sensitive to and breaking down the isms which often keep us apart. Ageism, Sexism, Racism. Meeting facilitation—developing non-manipulative skills. Decentralization—creating local alternatives. The Earth, our home, not their business.

FLOYD K. STEIN
Organisationen For Vedvarende Energi, Denmark

The energy question is more than a question about where all our energy will be coming from. It is a question about how we want to live in the future. We must choose that technology which fits that choice. It becomes more and more apparent that the most desirable direction is towards decentralization and local autonomy. And here the use of nuclear (or giant coal or giant solar for that matter) power is in direct opposition to this goal. In other words, local self-sufficiency, local democracy, cannot be combined with centralized energy logistics.

JUDY WILKS
Friends of the Earth, Australia

In Australia the threat to our future safety and sane energy future has currently to do more with the mining of uranium than with the construction of nuclear power plants. Australia at present has only one reactor. Despite continuous debate and demonstrations over the last five years, uranium mining is going ahead. And what's more, uranium mines are being successively sold to foreign interests. Our major trump card now has to do with the poor market outlook for Australian uranium. We are getting as much mileage out of this as we can from this debate, in the hope that if moral arguments are unconvincing, perhaps hip pocket ones will be. We are beginning to realize that in Australia at least, mass rallies are losing their support. Originally, say a few years ago, mass rallies were the major way of raising awareness. But the trouble was that we kept on having rallies and ignored many of the basic education tasks . . . for example the production of bilingual booklets and simple leaflets on the effects of radon gas in uranium mines, or radiation from nuclear power plants. Thus Friends of the Earth is attempting to tap the energies of the up-and-coming activists of the future who are still in school. We are spending a lot of time talking in schools, and in the coming summer holidays we are holding a "Sun Day" preceded by alternative technology workshops. However, we are still taking every opportunity to oppose uranium on the mining issue (as opposed to fighting it as part of a broader energy campaign). We are still heavily involved in lobbying trades unions and endeavouring to secure work bans in the uranium mines. We also attend the annual general meetings of large uranium mining companies as "Shareholders for Social Responsibility." At these meetings we ask questions about their activities and generally manage to get our point across.

JOHN HONTELEZ
Landelijk Energie Komitee, The Netherlands

If we talk about the anti-nuclear fight in the Netherlands, we talk about a fight against an international alliance of multinationals and governments. In Holland, the citizens, in majority, are against nuclear energy; this after years of intensive campaigns by environment-groups, left parties, churches, trade unions against nuclear energy. This is a unique situation, caused partly by the favorable energy situation in Holland (a big gas stock), the very little interest of the Dutch industry in atomic business, and the relative progressive view on environmental and military issues of the churches and trade unions. So, the political situation at the moment makes a further advance to a nuclear future very difficult. However, the European Community exercises great pressure to break the resistance and continue the nuclear programme, because of the uncertain oil situation and the need to strengthen the position of the French and German ("European," they say) atomic industry. Against this political and economic violence we can only succeed if we can break the left-right barrier and so rob the Dutch Conservative and Christian-Democratic servants of the European Community of their following in this point. Such a development can be started (is started, in fact) by: 1. Objective, scientific information to give about the dangers of nuclear energy, in particular about the Dutch projects. 2. To explain clearly the relations between the development of nuclear energy for electricity and the threatening of the world peace (proliferation), the police measures needed for safety, the lessening of democratic liberties that such measures cause, and the possibilities and dangers of atomic terrorism. 3. Prove that nuclear energy is not needed.. Pursue positive campaigns, for measures to save energy (but in a fundamental way, also in the production processes, also in the choice of what to produce), actions for sun energy and other environment-friendly ways to generate energy. To stress the realism of your solutions, to show your solutions in a real way. 4. Show that nuclear energy does not mean the end to dependence on the oil-exporting countries (which is a widespread misunderstanding). You cannot make petrol or chemicals of nuclear energy; besides, there grows a new dependency, on the uranium-exporting countries, a smaller, tighter group than OPEC. To succeed in this undertaking a very heterogeneous gathering of nuclear energy opposers must join their activities and means.

JOHN S. LAWLESS

Canadian Coalition for Nuclear Responsibility

It must be remembered that the Canadian economy is largely dominated by the U.S. Much of the investment capital available in Canada comes from the U.S. with the resultant flow of profits back south across the border. The Canadian government has always spent a lot of money with a view toward fostering homegrown industry. The CANDU reactor is only one example of an industrial project designed, built and marketed with huge amounts of government funding. In Canada at the present time there is no market for the CANDU. Ontario has a 40% excess over peak demand for electricity. Quebec, B.C. and Manitoba have lots of hydro-electric power. Alberta of course has oil, and Saskatchewan has coal and lots of uranium for export. Nova Scotia, Prince Edward Island and New Brunswick are embroiled in a battle, and are trying to extricate themselves from huge cost overruns for a reactor being built in New Brunswick, which some say will never operate due to technical and safety faults. Newfoundland has, in turn, discovered offshore oil. Canada has had little success in exporting reactors. This is for a variety of reasons, for which there is not enough time here to discuss. The industry will get continued support, however, as long as there is a potential market in the rest of the world. Opposition to the nuclear industry in Canada is focusing on export sales of nuclear hardware, software and on the mining and export of uranium, of which Canada has a bountiful supply. These deposits are mostly in British Columbia and Saskatchewan, but are also to be found in almost every other region in the country. The Saskatchewan government's response to the question of proliferation has been that if we don't sell uranium, someone else will. The federal government continues to export this uranium as a way out of a balance of payments problem. It will only be as a result of pressure brought to bear within Canada, in response to the severe environmental problems of mining that uranium exports will be halted. Short of a sharp decrease in demand for uranium and reactor technology in the rest of the world, the pro-renewable/anti-nuke forces will have to continue their efforts lobbying and educating the public and the politicians, with the belief that they will come to appreciate the false economies of the nuclear fuel cycle.

LARRY BOGART
Nuclear Hazards Information Center, U.S.A.

Based on the limited success citizens have had in halting nuclear power, it appears that the most effective method is door-to-door visits to friends and neighbors followed by a multi-sponsored town assembly at which prominent speakers, preferably a woman and a man, present the case against nuclear power. This approach is most productive when it follows hard on the heels of a utility announcement about possible sites for a nuclear reactor. For many years nuclear opponents exhausted their resources trying to deny licenses by intervening in the regulatory process, which, of course, is rigged. To date it has also been unproductive to lobby Congress and put pressure on the Executive and federal agencies, but increasingly citizens are getting through to township, county and state governing bodies. In time this promises to dry up sites for nukes and repositories for storage of nuclear waste. With the growing strength of the movement, a unified drive to change the composition of Congress might for the first time be fruitful in the 1980's.

PETRA KELLY

Die Grunen (The Green), Germany

These times are *bad* times—the clock of the Bulletin of the Atomic Scientists is moving closer and closer to midnight—Holocaust—when perhaps the earth will be burned and scorched—by the China syndrome or by a few Trident or Polaris missiles "accidentally" or in a planned way released from their silos. Although we, the worldwide growing nonviolent opposition (parliamentary and outer-parliamentary) have won many battles in courtrooms, on nuclear building sites and within electional processes, we are at the same time *unable* to stop for one moment the nuclear and chemical multinationals, the generals and guards and police forces, the irresponsible governments all around us, and the immoral unconcerned scientists and technicians working on death machinery like uranium bullets and silicon chips and fast breeder reactors are continuing to press ahead. Against life—for death. While the future of nuclear power is being called fully into question in the USA, and while Austria has defeated in a referendum one year ago the pro-nuclear lobby, other countries, obsessed by their nuclear ego trips are forging ahead. France wants to generate half of its electricity with the atom by 1985. The government is working to complete 35 further nuclear projects, including the fast breeder at Malville. In addition 20 atomic plants are on the drawing board at 28.3 billion dollars. France, in competing with the Federal Republic of Germany, aims to seize leadership in nuclear technology and so it sells nuclear plants to South Africa and Spain and it ignores the rise in childhood leukemias and cancer as is the case near La Hague. Scandal-ridden La Hague. Furthermore, it ignores the French ecological movement which gained nearly 4.5% of the vote last June. The 5% electoral restrictive clause introduced in France took away their seats in the European Parliament. In Germany there are 13 operating nuclear plants and work is nearing completion on three more. Nine are under construction and 11 still planned. Nevertheless, we have within the ecological movement halted construction on several plants by court injunction and we have been in some regions successful with an "electricity boycott" (deducting amount used for nuclear development from electricity bills and paying this into a special alternative energy fund).

We have tried to unite various movements, including the women's, the environmental, the peace and the anti-military groups, and have attempted to be present everywhere—on the nuclear sites, at the drilling holes in Gorleben, in the courtrooms, inside the city councils and regional Parliaments—the image of "flea in the fur" as a multi-strategy. We campaigned as the Green List for the European elections—a list of nearly 50% women and 50% men from different movements, each bringing with them their particular struggle and impact in the movement against a militarised and nuclearised world. We are still in the embryonic stage, but were able to get 3.2% during the European elections, and had major successes in the communal elections this past fall and in Bremen where we now have four fellow activists in the regional parliament. All this as part of the strategy—elections not as an *end,* but as *one* of the many, many *means* to fight for LIFE, for survival of the human species. All this, while our social democratic liberal government trades atomic "know-how" and nuclear hardware and uranium with military dictatorships like Brazil, with racist regimes like South Africa and assists in the exploitation of Australian Aboriginies and Navajo Indians because of the uranium race, the uranium crunch. These *are* bad times—for we still *do not have enough* allies in the trade union movements in Western Europe. There are a few courageous ones like Irish Transport and General Workers' Union which has, under the leadership of John F. Carroll, voted unanimously to oppose any future nuclear projects and will "block" any nuclear power station . . . There are courageous German trade unionists with a working group called LIFE and there are those who oppose any plans to expand the nuclear industry in France and in Holland. But we are still not enough. These are *bad* times—times in which employees from KWU, Germany's only reactor construction firm, have been informed they will be immediately fired if they engage in anti-nuclear opposition. Workers and employees of the giant firm must sign an agreement which forbids them to be members of a party or organization with an anti-nuclear position, to support anti-nuclear activities or even to have contact with them. Wives or husbands also have to declare that they will not in any way criticize the KWU firm.

We all know the interconnection between civilian and military nuclear power development–Siamese twins–in fact, there would never have been the so-called "peaceful" uses were there not the military reactors . . . . But now, a study by 66 nations and five international organizations suggest "that a dramatic increase in the availability of weapons grade, bomb-grade nuclear material is *inevitable.*" The report says that as many as 1,000 new nuclear plants may be built around the world by the year 2000 and concludes that there are *no* technical means of preventing this from increasing the risk of the spread of nuclear weapons. The report furthermore forecasts as much as *twelve fold* increase in the use of nuclear power in the non-Communist world by 2000. Worst of all, while forecasting a surge in existing types of nuclear plants, the report also states that the "development of a significant number of fast breeder reactors" could occur by 2000. Mrs. Thatcher in activating a new reprocessing plant at Counray has just revealed how much she believes in fast breeders. This is in line with thinking within the official European Community circles since the Community continues to spout out false wisdom about gaining energy independence through nuclear power, while all along uranium prices are skyrocketting and while uranium supplies prove unstable and will in the next two decades run out. Thus, the fast breeder, a reactor which produces more plutonium than it uses up, will be built everywhere they can get away with it, posing then even greater threats and reducing, in fact, cancelling all attempts to stem the spread of nuclear weapons. It will, of course, increase the practice of extracting plutonium, known as nuclear reprocessing, since there are and will be no honest solutions for safe eternal disposal of radioactive wastes. At least, what this recent international report by the International Nuclear Fuel Cycle Evaluation Conference has done is to *admit* fully and openly that civilian uses of nuclear energy can and do contribute to weapons proliferation.

And when one imagines that bombs of the size like the one dropped on Hiroshima are today only called "small tactical weapons," we must realize how close we are to total societal insanity. I am pleased that in other countries, and even in Eastern Europe, there are massive protests to stall, hinder, stop nuclear development. Spanish plans to foist nuclear plants on the poorest regions may well backfire. There is much resentment at the fact that Spain's poorest region, with little industry and backward agriculture, is providing sites for nuclear power plants whose output will be consumed in industrial and urban centres. This is the usual tactic of the nuclear lobby–trying to seduce such regions with empty promises of jobs, industrial development and a new quality of life . . . In Sweden, plans are under way for a national referendum that could mean the dismantling of the country's six nuclear reactors. And even in the Eastern bloc, reservations are being voiced about the ambitious programmes. In Yugoslavia, there are first victories for the anti-nuclear lobby. In 1978, under strong pressure, the Local Assembly of Zadar removed the nuclear plant from its long-term energy plan. Fear of radiation and a nuclear mishap and the impression thousands of Yugoslavian migrant workers had brought home led to the rejection of plans approved by the ruling Communist League. In the last two weeks, the Communist Party's theoretical organ (Kommunist) revealed anxiety on the possible harmful effects of nuclear power on the environment and revealed the problems of nuclear waste storage, of siting, etc. I should point out that in 1979, the Yugoslavian government dropped the plan to build a nuclear power plant off Zadar.

Great Britain, user of atomic power since the mid 1950's, now has 19 nuclear power stations and Prime Minister Thatcher's Conservative government is expected to unveil plans to build at least 20 new reactors by the year 2000. I cannot appeal to Mrs. Thatcher's instincts as a woman, as someone who should put a *stop* to instead of furthering the nuclear power program—for she has adjusted to and emulated far too much the patriarchal values of a power hungry, competitive society—where we build up multi-billion eternally lethal nuclear energy systems just in order to *boil water*—in order to produce electricity—in order to boil water again. Since taking office, Mrs. Thatcher has placed her priorities on devoting money to modernizing the British military and in giving generous pay rises to the armed forces and increasing total defense spending next year by over 3%—all this while the economy is ailing, while there is a lack of hospitals and day-care centres, while there are not enough kidney dialysis machines, while women are continually underpaid, while bad housing drives large poor families into desperation. And Mrs. Thatcher is also a leading outspoken proponent of the modernisation of NATO's nuclear weapons and has agreed to have new nuclear missiles deployed at bases in eastern England. And the Tory government is moving towards a police state—as are all nuclear nations. Nuclear energy supposedly "protecting" the public is a good excuse. The proposed new Protection of Official Information Bill in the U.K., I am told, could wholly and effectively destroy press and personal freedom in key areas of policy. The bill does not confine itself to "disclosure" in conversation or by publication. It gives officials powers to tour the country if they wish, demanding that books, documents containing such "protected" information be surrendered.

Already the infamous Trilateral Commission has produced a study entitled *The Crisis in Democracy* and has stated that the development of democracy is becoming a danger for the very future of Western societies—necessitating restraint on democracy and fundamental liberties. It is for ethical, moral, health, political, economic and social reasons that we must continue to oppose non-violently, but effectively the nuclear power programme. We will have to resort to many strategies—such as blocking uranium shipments, creating new radical forms of adult education such as the folk high school in Wyhl, occupying nuclear sites and building adventure playgrounds on top of them or planting trees and flowers as at Gorleben and Larzac. We will also need to tie up projects in court—at national and European levels and we must make choices as to how to gain seats within local and regional councils—for there too, our voices must be heard . . . In a recent court case where two German defendants had witheld 10% of their electricity bills, the Magistrate Ms. Adelheid Kiefnere ruled *in their favor.* The magistrate declared that nuclear producers were operating on a "trial and error" basis which represented a public danger. The utility has a duty under law to ensure that the public is protected. Therefore, the judge argued, the consumers were right to refuse payment. Because of "trial and error"—because plutonium is "thalidomide forever" and will make states into totalitarian systems, we rise up—and demand that the creativity and imagination of human beings be diverted from self-destruction and destruction of others to soft, renewable, decentralized energy systems and to an ecological society—where one *is* and does not only *have.*

JOHN GOFMAN
Committee for Nuclear Responsibility, U.S.A.

If President Carter, the Congress, and the nuclear industry believed in the safety of nuclear power, they would all immediately demand repeal of the Price-Anderson Act—that pernicious law which absolves the utility industry of liability for the causation of your death and of your property damage in a nuclear accident. The nuclear menace would end in 24 hours after repeal of the Price-Anderson Act, because utilities will risk your life but not their dollars. They have no confidence in nuclear safety. Let President Carter know that 100 million Americans demand his support for immediate repeal of the Price-Anderson Act. I hope 100 million Americans tell their congressmen and senators to repeal Price-Anderson this week. Congressman Ted Weiss has a bill on this in the House already. It is important that the required action not be subverted into raising the public subsidy for utilities through even more Price-Anderson coverage. We need the repeal of that vicious law, and placement of unlimited liability back on the utilities where it belongs, since that is the only way to get them to act responsibly with respect to public health and safety. Can you imagine a worse law? This is a law which gives you, the American people, the privilege of paying the salaries of bureaucrats who arrange your death-warrants, and the additional privilege of paying the profits of the companies which act as your executioners. Repeal of that law would be enough to end the nuclear power problem.

VALERI NERVI
Amici Della Terra, Italy

The problem is one of the most difficult to cope with but even if we know there isn't an easy solution, here is our strategy. We think that people must be continuously fed with good information. Mass media, on the contrary, tries to make people forget their own responsibilities and to leave the solution of energy problems to technicians and to politicians as "the experts." We think that nuclear power can be stopped only by the people's will. And, they must be convinced that their future will be as safe and comfortable as nowdays. To this aim, we organize high-level conferences with Italian and foreign scientists to give serious, scientific support to our fight, such as the conference held in Rome in May, 1979 with Amory Lovins. We publish magazines and books choosing and translating the best international studies on energy, and we publish our own bulletin. We have requested a referendum against Law 393, which regulates nuclear site choices in Italy. We have charged MHB Technical Associates, an American firm specialized in nuclear safety problems, to do a complete and scientific inquiry about the safety of the Italian nuclear plants. The results of this study, which is similar to a study committed to MHB by the Swedish government, will be presented in several conferences to the public and to the local authorities.

CONNIE HOGARTH
Women's International League For Peace and Freedom, U.S.A.

I believe very deeply that nuclear power and nuclear weapons will not be stopped until the people want them stopped—until people realize that their lives and their very future depend on whether we stop this technology. I personally feel a commitment to help people feel empowered to make a change, which requires that one believe there is an importance, an expertise in being a human being. We have been conned by experts and expertise bringing us close to any number of disasters: from PCB's to the atom bomb to nuclear power plants to plutonium. It is time people realize that they have the judgement to determine what's best for them. In terms of implementation, once we have a sense of our power it is like the spokes on a wheel. Although not all of us are ready to lay our bodies on the line, not all of us are taken with lobbying our representatives, not all of us can write letters, not all of us can do public speaking, not all of us can march–we all can do something. And like the spokes of a wheel, even though it may seem we're 90 degrees apart, if each of our spokes are directed towards the hub, which means survival, turning away from nuclear technology and exposing those who profit from it, no matter how diverse our methods, we are strengthened and we are part of the power in moving that wheel. I think we have to use every non-violent means at our disposal to make the changes that are to guarantee a future.

WILLIAM WINPISINGER
Labor Leader, U.S.A.

In 1976, Jimmy Carter campaigned, on a contract with the people, to break up the energy conspiracy. Now we know he was a double agent for the Corporate State. First, he lifted price controls from natural gas–a $50 billion windfall to oil and gas producers. And a 40 per cent increase in home heating and cooking bills. Second, he's taking price controls off domestic crude oil, gasoline and home heating oil–another $80 billion to $300 billion swindle at the expense of workers, consumers, senior citizens, the disabled, and unemployed. Third, Carter is putting our future energy eggs in the nuclear power basket. He went to Three Mile Island to show where his head is. He's cut the solar energy budget, stifled conversion to coal, poor mouthed conservation, increased the nuclear budget, and come up with expensive synfuels to compete with nuclear in a contract to determine which is the most lethal and hazardous. Only Edward Teller, America's mad scientist, could write a worse energy scenario. Nuclear energy is Kurt Vonnegut's Ice Nine. It is the Agent Orange of the energy industry. It is the most inflationary means of electric power. It is, as Barry Commoner has said, a dumb way to boil water. The cost of a nuclear power plant is two-and-a-half times the cost of coal conversion. And think about this: the domestic uranium supply is projected to be in shortfall by the end of the 1980's. Then we'll have to go to foreign suppliers to find uranium. Carter's mania for nuclear energy means we're trading OPEC for UPEC (Uranium Producers Export

Cartel). Australia and Canada may not be too bad, but how about South Africa? That's who we'll be depending on to fuel our reactors.

And in the meantime, guess who owns the uranium resources of this country? The same corporations that have brought us the natural gas swindle, the phony oil shortage, the gasoline lines, and the heating oil rip-off. What we have to tell Jimmy Carter and the nuclear stooges in Congress is we can't afford to place the nation's energy future in the sticky palms of those who've pulled off the biggest heist in the history of mankind. Nuclear energy isn't safe at any cost. Nuclear waste can't be safely dumped anywhere. The energy conspiracy has squandered billions of taxpayer dollars subsidizing nuclear power. And all we have to show for it is the cancer connection and a 300,000 year curse on the earth and future generations. If we committed our tax dollars to developing solar, low-head hydro, biomass, co-generation, and the rest, we'd be off the OPEC cartel hook, and out of the China syndrome, too

There is no such thing as low-level radiation. There is only deadly radiation. You can't see, feel, taste, or hear it. But it's always there, around you. And the damned stuff is poison. The machinist union physician, Dr. Thomas Mancuso, attempted to expose the low-level radiation cover up ten years ago. Industry and government regulatory agencies conspired to suppress his findings. Dr. Mancuso was forced to resign. Today, nuclear power is offering to hire him back. He's not selling. We only have to know this about all nuclear energy—whether it's power plants or weapons—the first line of victims of radiation-induced cancer and leukemia are the workers who handle the stuff. They're five times as apt to get cancer as the rest of the working population. Workers took the first lethal radiation baths at Three Mile. Workers who have to haul and transport nuclear waste and other radioactive materials are the first to be dosed in case of collision or accident on the highways and byways. Workers who handle radioactive materials in warehouses, air terminals, and on the shipping docks, are constantly exposed to radiation ooze. Nuclear power and radiation constitute the cancer connection in the nuclear labor force.

But the Big Energy Conspiracy tells us that without nuclear power plants all over we won't have enough energy. They say, we have to have nuclear power because we need the jobs. More energy blackmail. Nuclear blackmail. Job blackmail. Look who's telling us they are concerned about our jobs. The $100 billion nuclear industry roll call—Gulf, Exxon, Mobil, Kerr-McGee, Westinghouse, G.E., Babcock & Wilcox, Stone & Webster, Allied Chemical, Dow, DuPont, Rockwell, Brown & Root, Flour, Bechtel, Met Ed, Con Ed, Com Ed, etc. The

Atomic Industrial Forum, National Association of Manufacturers, Business Roundtable, Chamber of Commerce, Edison Electric Institue. They're worried about our jobs? Like they worry about National Health Insurance. Like they worry about the Occupational Safety and Health Law. Like they worried about the Consumer Protection Agency, labor law reform, and construction workers' David-Bacon law. Like they worry about their workers whenever they try to organize unions. That kind of worry, we can do without. Job blackmail, we can do without. We can no longer expect Jimmy Carter to answer anyone's roll call, except Big Energy's. But before his hair falls out, he could do humanity one good turn, and ban the neutron bomb—the ultimate Corporate State weapon: it kills people and preserves property. And there's the connection between nuclear power and weapons and war. You can generate electricity and make A-bombs, too. India did it, built a nuclear plant and, undetected by the International Inspection Agency, built an A-bomb using the uranium from the power plant fuel rods. If India can do it, so can any other country. Nuclear power means nuclear proliferation around the globe. Ice Nine.

If the Corporate State will draw up plans to go to war for oil in the Persian Gulf—go to war for OPEC—it will go to war for uranium in South Africa—go to war for UPEC. And now they're telling us we are not only in a nuclear weapons race with Russia—we're in a nuclear power race, too. We have to stop this nuclear madness. Stop the lying. Stop the cover ups. Stop the price rip-offs. Stop the Energy Conspiracy. Stop Big Oil. It is time to put the people in charge of this nation's energy policy and development. Put back price controls. Put the moratorium on nuclear. Tell it to Congress. Tell it to the White House. Meantime, follow that time-honored trade union principle: agitate, aggravate, demonstrate, educate, organize and mobilize. And remember—Nuclear destroys. Solar employs.

LEAH WARN
World Information Service on Energy, The Netherlands

In one word, "ACTION."

RECOMMENDED READING AND VIEWING

OVERVIEW

Nuclear Power: The Unviable Option, John J. Berger, Ramparts Press, Palo Alto, California, 1976.

No Nukes: everyone's guide to nuclear power, Anna Gyorgy and friends, South End Press, Boston, 1979.

The Menace of Atomic Energy, Ralph Nader and John Abbotts, W.W. Norton & Company, New York, 1977.

The Silent Bomb, A Guide to the Nuclear Energy Controversy, Edited by Peter Faulkner, Random House, New York, 1977.

Radioactive Contamination, Virginia Brodine, Harcourt Brace Jovanovich, New York, 1975.

Accidents Will Happen, The Case against Nuclear Power, Edited by Lee Stephenson and George R. Zachar, Harper & Row, New York, 1979.

Nuclear Power, Walter Patterson, Penguin Books, England, 1976.

Perils of the Peaceful Atom, The Myth of Safe Nuclear Power Plants, Richard Curtis and Elizabeth Hogan, Ballantine Books, New York, 1970.

The Anti-Nuclear Handbook, Stephen Croall and Kaianders, Pantheon Books, New York, 1978.

The Electric War, Sheldon Novick, Sierra Club Books, San Francisco, 1976.

ACCIDENT HAZARDS

The Accident Hazards of Nuclear Power Plants, Richard E. Webb, The University of Massachusetts Press, Amherst, Massachusetts, 1976.

We Almost Lost Detroit, John G. Fuller, Reader's Digest Press, New York, 1975.

The Nuclear Fuel Cycle, The Union of Concerned Scientists, The MIT Press, Cambridge, Mass., 1974.

Nuclear Disaster In The Urals, Zhores A. Medvedev, W.W. Norton & Comapny, New York, 1979.

MEDICAL CONSEQUENCES

Nuclear Madness: What You can Do!, Dr. Helen Caldicott, Autumn Press, Brookline, Mass., 1978.

Poisoned Power, The Case Against Nuclear Power Plants, John W. Gofman and Arthur R. Tamplin, Rodale Press, Emmaus, Pa., 1971.

Malignant Neglect, Environmental Defense Fund and Robert H. Boyle, Alfred A. Knopf, New York, 1979.

Low-Level Radiation, Ernest J. Sternglass, Ballantine Books, New York, 1972.

The Politics of Cancer, Dr. Samuel S. Epstein, Sierra Club Books, San Francisco, 1978.

Are You Radioactive?, Linda A. Clark, Pyramind Books, New York, 1973.

THE ALTERNATIVES

Soft Energy Paths: Toward a Durable Peace, Amory B. Lovins, Harper & Row, New York, 1977.

Non-Nuclear Futures: The Case for an Ethical Energy Strategy, Amory B. Lovins and John H. Price, Ballinger Publishing, Cambridge, Mass., 1975.

The Sun Betrayed, A Report on the Corporate Seizure of U.S. Solar Energy Development, Ray Reece, South End Press, Boston, 1979.

World Energy Strategies, Amory Lovins, Ballinger Publishing, Cambridge, Mass., 1975.

Rays of Hope, Denis Hayes, W.W. Norton & Company, New York, 1977.

Energy for Survival, The Alternative to Extinction, Wilson Clark, Anchor Books, Garden City, N.Y. 1975.

It's In Your Power, The Concerned Energy Consumer's Survival Kit, Stuart Diamond and Paul S. Lorris, Rawson, Wade Publishers, New York, 1978.

POLITICS OF ENERGY

The Politics of Nuclear Power, Dave Elliott, Pluto Press, London, 1978.

The New Tyranny, Robert Jungk, Grosset & Dunlap, New York, 1979.

Energy War: Reports from the Front, Harvey Wasserman, Lawrence Hill, Westport, Conn., 1979.

Light Water, How the Nuclear Dream Dissolved, Irvin C. Bupp and Jean-Claude Derian, Basic Books, New York, 1978.

The Nuclear-Power Rebellion, Citizens vs. the Atomic Industrial Establishment, Rochard S. Lewis, The Viking Press, New York, 1972.

The Politics of Energy, Barry Commoner, Alfred A. Knopf, New York, 1979.

The History of the Standard Oil Company, Ida Tarbell, Peter Smith, Gloucester, Mass., 1963.

ECONOMICS

Nuclear Power Costs, U.S. House of Representatives Committee on Government Operations, Washington, D.C., 1978.

Nuclear Power: The Bargain We Can't Afford, Environmental Action Foundation, Washington.

ALSO

The Poverty of Power, Barry Commoner, Bantam, New York, 1977.

The Closing Circle, Barry Commoner, Bantam, New York, 1974.

Irrevy, An Irreverent Illustrated View of Nuclear Power, John Gofman, Committee for Nuclear Responsibility, San Francisco, 1979.

WORTH WRITING TO FOR VARIOUS BOOKLETS

Committee for Nuclear Responsibility, P.O. Box 11207, San Francisco, CA 94101

National Council of Churches Energy Project, 465 Riverside Drive, New York, N.Y. 10027

Physicians For Social Responsibility, P.O. Box 295, Cambridge, Mass. 02238.

Union of Concerned Scientists, 1208 Massachusetts Ave., Cambridge, Mass. 02138

Nuclear Information and Resource Service, 1536 16th St., NW, Washington, D.C. 20036

Environmental Action Reprint Service, 2239 East Colfax, Denver, Colorado 80206

PERIODICALS

Critical Mass Journal, P.O. Box 1538, Washington, D.C. 20013

Environmental Action, 1346 Connecticut Ave., NW, Washington, D.C. 20036

New Roots, P.O. Box 459, Amherst, Mass. 01002

Not Man Apart, Friends of the Earth, 124 Spear St., San Francisco, Calif., 94105

Nuclear Opponents, Box 285, Allendale, N.J. 07401

RAIN, 2270 N.W. Irving, Portland, Ore. 97210

FILMS AND VIDEO

Available from Physicians For Social Responsibility, Box 295, Cambridge, Mass. 02238

Paul Jacobs and the Nuclear Gang, 16mm.

The Medical Implications of Nuclear Energy, 3/4 video.

Danger! Radioactive Waste, 16mm.

Decision at Rocky Flats: A Question of Trespass, 3/4 video.

Clouds of Doubt, 3/4 video.

Available from Green Mountain Post Films, Box 177, Montague, Mass. 01351

Save the Planet, 16 mm.

The Last Resort, 16mm and 3/4 video.
Lovejoy's Nuclear War, 16mm and 3/4 video.
More Nuclear Power Stations, 16mm.
Sentenced to Success, 16mm.
Better Active Today than Radioactive Tomorrow, 16mm.
Nuclear Reaction in Whyl, 16mm.
Radiation and Health, 16mm.
Training for Nonviolence, 16mm.
The Accident, 16mm.
Danish Energy, 16mm.
Uranium Mining in Australia, color slides.
The Atom and Eve, 16mm.
Early Warnings, 16mm.

ACKNOWLEDGMENTS

With grateful appreciation for the help given to me on this book by, most of all, my wife, Janet Grossman, who worked with me every inch of the way; Lorna Salzman, Mid-Atlantic representative of Friends of the Earth, who made me aware of nuclear power many years ago; Dr. Richard Webb, nuclear scientist and engineer whose immense knowledge is equaled by his conscience; writer Val Schaffner, who assisted with Chapter Three, and my journalistic associate, John Rather, who assisted with Chapter Eight; Julia Ludmer, my editor and Eric Salzman, for his editing assistance; Dr. Stephen Sigler; Arnold Hoffmann; Maggy Simony, Glenferrie Associates; Van Howell, Richard Partington and Leslie Barker of Friends of the Earth who assisted with Chapters Four and Six; and Ingrid Arensen of the SHAD Alliance who assisted with Chapter Nine. To the U.S. Freedom of Information Act. And to courageous publishers, Dr. Martin and Judith Shepard.

We appreciate the cooperation of:

Alex Cockburn: Excerpt from "Tales of Power, A Special Section on Nuclear Power, The No Nukes Movement and Alternative Currents," by Alex Cockburn and James Ridgeway, in *The Village Voice,* May 7, 1979. Reprinted by permission of Alex Cockburn.

Alfred A. Knopf, Inc.: Excerpt from *Malignant Neglect* by the Environmental Defense Fund and Robert H. Boyle. Copyright 1979 by the Environmental Defense Fund and Robert H. Boyle. Reprinted by permission of Alfred A. Knopf, Inc.

Autumn Press: Excerpts from *Nuclear Madness: What You Can Do!* By permission from *Nuclear Madness: What You Can Do!* by Dr. Helen Caldicott. Copyright 1978 by Helen Caldicott. Reprinted by permission of Autumn Press, Brookline, Massachusetts.

Estate of Albert Einstein: Excerpts from *Out Of My Later Years* by Albert Einstein. Copyright 1950 by the Philosophical Library. Reprinted by permission of the Estate of Albert Einstein.

Environmentalists for Full Employment: Excerpts from *Jobs and Energy* by Environmentalists for Full Employment. Copyright 1977 by Environmentalists for Full Employment. Reprinted by permission of Environmentalists for Full Employment.

Friends of the Earth: Excerpts from *Soft Energy Paths: Toward a Durable Peace* by Amory Lovins. Copyright 1977 by Friends of the Earth. Reprinted by permission of Friends of the Earth.

Lyle Stuart: Excerpt from *The Rockefeller Syndrome* by Ferdinand Lundberg. Copyright 1975 by Ferdinand Lundberg. Printed by arrangement with Lyle Stuart.

Prentice-Hall, Inc.: Excerpts from the book *City of Fire* by James W. Kunetka. Copyright 1978 by James W. Kunetka. Published by Prentice-Hall, Inc., Englewood Cliffs, N.J. 07632. Reprinted by permission of Prentice-Hall, Inc.

Princeton University Press: Excerpts from *Change, Hope and The Bomb* by David E. Lilienthal. Reprinted by permission of Princeton University Press.

Ramparts Press: Excerpts from *Nuclear Power: The Unviable Option* by John J. Berger. Copyright Ramparts Press, Palo Alto, California. Reprinted by permission of Ramparts Press.

Random House, Inc: Excerpt from *Energy Future: Report Of The Energy Project At The Harvard Business School,* edited by Robert Stobaugh and Daniel Yergin. Copyright 1979 by Robert Stobaugh and Daniel Yergin. Reprinted by permission of Random House, Inc.

Rawson, Wade Publishers, Inc.: Excerpts From *It's In Your Power* by Stuart Diamond and Paul S. Lorris. Copyright 1978 by Stuart Diamond and Paul S. Lorris. Reprinted by permission of Rawson, Wade Publishers.

Rodale Press: Excerpts from *Poisoned Power.* Reprinted from *Poisoned Power.* Copyright 1971 by John W. Gofman and Arthur R. Tamplin. Revised 1979. Permission granted by Rodale Press, Inc. Emmaus, PA 18049.

Ruttenberg, Friedman, Kilgallon, Gutchess & Associates, Inc.: Excerpts from *The American Oil Industry, A Failure Of Anti-Trust Policy.* Prepared for the Marine Engineers' Beneficial Association by Stanley H. Ruttenberg and Associates, Inc. Excerpts from *The Energy Cartel, Big Oil vs The Public Interest* by Ruttenberg, Friedman, Kilgallon, Gutchess & Associates, Inc. Copyright 1974, 1975. Reprinted by permission of Ruttenberg, Friedman, Kilgallon, Gutchess & Associates, Inc.

Sierra Club Books: Excerpt from *The Politics of Cancer* by Dr. Samuel Epstein, M.D. Copyright 1978 by Samuel S. Epstein. Reprinted by permission of Sierra Club Books.

The Trilateral Commission: Excerpts from *The Crisis of Democracy* by Michael J. Crozier, Samuel P. Huntington, Joji Watankuki. Copyright 1975 by The Trilateral Commission. Reprinted by permission of The Trilateral Commission.

Union of Concerned Scientists: Excerpts from *Looking But Not Seeing* by Lawrence S. Tye. Copyright 1979 by the Union of Concerned Scientists. Reprinted by permission of the Union of Concerned Scientists.

Union of Concerned Scientists: Excerpt from *Nuclear Fuel Cycle* by the Union of Concerned Scientists. Copyright 1974 by the Union of Concerned Scientists. Reprinted by permission of the Union of Concerned Scientists.

Viking Penguin Inc.: Excerpts from *Between Two Ages* by Zbigniew Brzezinski. Copyright 1970 by Zbigniew Brzezinski. Reprinted by permission of Viking Penguin Inc.

Viking Penguin Inc.: Excerpts from *The Nuclear Power Rebellion* by Richard S. Lewis. Copyright 1972 by Richard S. Lewis. Reprinted by permission of Viking Penguin Inc.

William Morrow & Company, Inc.: Excerpt from *The People's Almanac #2* by David Wallechinsky and Irving Wallace. Copyright 1978 by David Wallechinsky and Irving Wallace. Reprinted by permission of William Morrow & Company, Inc.

W.W. Norton & Company, Inc.: Excerpts from *Rays of Hope* by Denis Hayes. Copyright 1977 by Worldwatch Institute. Published by W.W. Norton & Company, Inc. Reprinted by permission of W.W. Norton & Company, Inc.

INDEX